河陽鄉村辭典

卓旻 著

中国建筑工业出版社

图书在版编目（CIP）数据

河阳乡村辞典 / 卓旻著. -- 北京 ：中国建筑工业
出版社，2022.6
ISBN 978-7-112-27455-0

Ⅰ．①河… Ⅱ．①卓… Ⅲ．①村落－古建筑－缙云县
－词典 Ⅳ．①TU-092.2

中国版本图书馆CIP数据核字（2022）第107137号

责任编辑：唐旭 吴绫
文字编辑：李东禧 孙硕
责任校对：王烨

河阳乡村辞典

卓旻 著

*

中国建筑工业出版社出版、发行（北京海淀三里河路9号）

各地新华书店、建筑书店经销

北京中科印刷有限公司印刷

*

开本：889毫米×1194毫米 1/20 印张：11⅗ 字数：289千字

2022年7月第一版 2022年7月第一次印刷

定价：58.00元

ISBN 978-7-112-27455-0

（39523）

DICTIONARY

of

RURAL

HEYANG

河阳乡村辞典

序

PREFACE

　　真正开始计划做一些乡村相关的研究始于大约十年前。那时在上海正在为世博会的一些场馆做策划和设计工作，上海世博会的主题是"城市让生活更美好"。上海世博会的主题映射的是当时城市化快速发展的阶段。从语言逻辑来看，这个主题很容易地就引申出这样的问题：那乡村生活就很糟糕吗？如果到街头采访那些迁居到城市的乡村人，这个问题的答案很明显，乡村好谁还会离开乡村啊？但是说完这话之后往往会跟一个遗憾的评论，其实老家也蛮好的。所以这个问题变得不是那么容易回答，或者说，乡村的问题远比想象中复杂。

　　乡村的议题在十多年前不像现在这么广泛受人关注。乡村建设在一些敏锐的人眼中更像是一种艺术观念，因为它天然具有观念艺术的重要特点。乡村意味着边缘，意味着对农民这一弱势群体的关照，在拥抱城市化的时代中具有很强的批判意识。而且乡村提供了一个单纯而朴素的巨大背景，在乡村中的任何创作或建设，小如乡村图书馆或是乡村厕所，其所蕴含的些许观念都可以被这个背景烘托出一种宏大感。但是如果乡村建设只是停留在一个艺术概念的层面，那是缺乏生命力的，从伦理角度而言也是有缺陷的。新时代的乡村建设固然需要新鲜的创意，但更重要的是去真正了解中国乡村的过去和现在，了解村民们的切身需求和感受。

5

这十年间，带着学生走了很多浙江的乡下角落进行田野调查，让学生切实感受时代变革中的乡村。设计师谈乡村问题时往往偏于保护的议题，而疏于发展。所以自己也零零落落地对乡村的结构、原型、现象、现代性等问题做了一些认真的思考，把这些想法整理到一起好像也能串成一条基于语言学的乡村研究线索。这些思考没有沉溺于过去，毕竟现代性已经渗透到所有人的生活当中，尊古而陋今是不可取的。

浙江省缙云县的河阳古村提供了一个很好的乡村语言学的实验样本。自 2014 年始，我几乎每年都带学生来到这里进行教学实验。河阳是一个省级保护村落，但相比其他很多被开发为旅游景点的古村，这里还是非常具有生活气息。而且它也不是孤立的，在流经它的大溪的上下游还有大小不同、形态格局各有特点的自然村落。这些村子正经历着和中国任何一个地方的普通乡村一样的变化，古村传统的延续和外部世界的渗透在这里互相抵抗和妥协，它们合在一起构成一片最为真实的当代乡村场景。

所谓的河阳乡村辞典，是找到那些因时代的侵蚀而解体的空间和符号的碎片，将它们看作是辞典里的一个个词条来进行释义并探讨这些词条的心灵意向和原始潜能。这当然不是一本传统意义上的辞典，在这些乡村词条的释义过程当中，摄影、文本、速写 —— 这些不同的书写方式形成一种新的各类文本自由对话的互文语境，以避免历史意义被固定在能指之中而被定格为某种陈词滥调。每一个词条作为符号而言都不是固定的，而是有待于被使用者重新书写来生成新的意义。

缘起性空。在这个已被现代性包裹的土地上，对于这些乡村碎片的观察、解释、演绎，或许也不过只是一种幻象，因缘而生而已。

卓旻

2021 年 4 月 17 日

目 录

第一章

乡村的思考

你这样的匆忙，你可有什么事？
我要看还有没有我的家乡在；
我要走了，我要回到望天湖边去。
我要访问如今那里还有没有白波翻在湖中心，
绿波翻在秧田里，
有没有麻雀在水竹枝头耍武艺？
先生，先生，世界是这样的新奇，
你不在这里遨游，偏要哪里去？
我要探访我的家乡，我有我的心事；
我要看孵卵的秧鸡可在秧林里，
泥上可还有鸽子的脚儿印"个"字，
神山上的白云一分钟里变几次，
可还有燕儿飞到人家堂上来报喜。
先生，先生，我劝你不要回家去；
世间只有远游的生活是自由的。
游子的心是风霜剥蚀的残碑，
碑上已经漶漫了家乡的字迹，
哦，我要回家去，我要赶紧回家去，
我要听门外的水车终日作鼍鸣，
再将家乡的音乐收入心房里。
先生，先生，你为什么要回家去？
世上有的是荣华，有的是智慧。
你不知道故乡有一个可爱的湖，
常年总有半边青天浸在湖水里，
湖岸上有兔儿在黄昏里觅粮食，
还有见了兔儿不要追的狗子，
我要看如今还有没有这种事。
先生、先生，我越加不能懂你了，
你到底，到底为什么要回家去？
我要看家乡的菱角还长几根刺，
我要看那里一根藕里还有几根丝，
我要看家乡还认识不认识我，
我要看坟山上添了几块新碑石，
我家后园里可还有开花的竹子。

闻一多·《故乡》

Rural Construction & National Salvation

—— Rural Experiment in the Republic of China Period

The countryside used to be the hometown of most people, a home interwoven with images of rice paddies, hall fronts, yellow dogs, water chestnuts, etc. For those who lived there, it was an idyllic garden filled with life; for those who were far from it, it was an unnamable container of nostalgia. But such a scene has really left away from us.

Regarding the decline of the Chinese countryside, it is not something that has only emerged recently, but really could be dated back to the Period of Republic of China a century ago. Liang Shumin, a famous Confucian scholar of that time once pointed out that "the destruction of the countryside in China in the last hundred years has been an absolute devastation, which has not been seen anywhere in the world." *(The Complete Works of Liang Shuming. 2005. P153)* He believed that the devastation had lasted for a century. It can be assumed that the decline of the Chinese countryside started since the collapse of the traditional Chinese society after the Opium War. So in terms of time, it has been nearly two hundred years till now.

During the period of Republic of China, scholars studied the Chinese rural issues and took out practices from different perspectives.

Tao Xingzhi and Yan Yangchu, upholding the ideals of enlightening the people and

乡建与救亡

—— 民国的乡建实验

　　乡村曾经是大多数人的故乡，是一处由秧田、堂前、黄狗、菱角等物象交织而成的家。对于生活在那里的人而言，那是弥漫着生活气息的田园；对于远离乡村的人而言，那又是不可名状的乡愁容器。但是这样的景象，实在地在离我们远去。

　　关于中国乡村的衰退问题，其实不仅是在当下，早在百年前的民国就已经是一个凸显的问题。民国大儒梁漱溟先生曾经指出"中国近百年来的乡村破坏，是一种绝对破坏，为世所仅见。"[1]梁漱溟当时即已认为中国乡村的破坏已有百年之久，可见中国乡村的衰败实则始肇于鸦片战争之后的整个传统中国社会的崩溃，所以要论时间的话，到现在也有近两百年了。

　　民国时期，不同的学者从不同的视角对中国乡村问题进行了研究和实践。

　　陶行知和晏阳初秉持启发民智、固民宁邦的理念，以乡村为重点大力推行平民教育。在二十世纪二十年代，晏阳初以一种传教士的精神，带着几十位海外留学归来的博士从北京来到河北定县进行乡村综合社会改造实验。这场当时被称为"博士

developing the nation, vigorously promoted civilian education with a focus on the countryside. In the 1920s, Yan Yangchu, in a missionary spirit, led dozens of doctors who returned from overseas from Beijing to Ding County in Hebei province, to conduct a comprehensive social transformation experiment of the countryside. This experiment known as "Association of Promoting Mass Education (APME)" experiment, lasted for ten years, aiming to combine civilian education with rural transformation. Through his research in Ding County, Yan Yangchu categorized Chinese rural issues into four roots: stupidness, poverty, feebleness, and selfishness. He advocated literature and arts education to dissolve stupidness and cultivate civilian knowledge, vocational education to reduce poverty and increase productivity, health education to cure feebleness and build strong physique, and citizenship education to overcome selfishness and develop a spirit of unity. However, this missionary attitude overstated the dualistic antagonism between modernity and tradition.

When the industrialist Lu Zuofu witnessed the collapse of the small-scale peasant economy when confronted with the industrial competition in the Chinese countryside, he believed that the solution to the rural problem should start with the revitalization of industry, and "countryside should be quickly modernized" through the development of rural production and infrastructure construction. Lu Zuofu had a profound understanding of the social structure of the Chinese countryside. When discussing the twofold social life of the Chinese, he explained that "people often blame the Chinese for knowing only the family but not the society; in fact, the Chinese have only the family but no society, since the family is the society for the Chinese." *(Lu Zuofu. Selected Works of Lu Zuofu. Edited by Lin Yaolun, Xiong Fu. 1999. P323)* Based on this understanding, Lu Zuofu put forward the concept of building a modern collective life, hoping that a new rural social structure with society as the core could replace the traditional rural social structure that valued the family most. From the late 1920s, Lu Zuofu presided over and promoted the rural construction movement in more than thirty townships in the Three Gorges area of the Jialing River, with Beipei in Chongqing as the center. In Beipei, a township still in the medieval time, he began by building a new hot spring park to give local people an idea that fineness and harmony could be created by their own hands. He also founded normal schools, built library, museum, sports grounds, hospitals, newspaper, etc., to "make the Three Gorges of the Jialing River area as a productive area, a cultural area, and a tourist area." *(Lu Zuofu. Selected Works of Lu Zuofu. Edited by Lin Yaolun, Xiong Fu. 1999. P343)* He introduced a lot of production businesses in the rural construction in Beipei and further developed cultural and social public enterprise, hoping to bring material and spiritual civilization to the traditional countryside.

The sociologist Fei Xiaotong devoted much of his sociological fieldwork to the study of the Chinese countryside. In his early work "From the Soil: the Foundations of Chinese Society", he made a unique analysis of the "stupidness" and "selfishness" of the country people, and explored the vernacular nature of Chinese society. As for "stupidness", he argued that the "stupidness" of the country people was in fact the result of the formation of a more direct "special language" system inclined to understanding beyond the common writing system

下乡"的实验持续了十年之久，希望将平民教育与乡村改造相结合。通过在定县的调研，晏阳初认为中国农村普遍存在着"愚、贫、弱、私"这四大病根，他提倡以文艺教育化"愚"，培养平民的知识力；以生计教育化"贫"，培养生产力；以卫生教育化"弱"，培养强健力；以公民教育化"私"，培养团结力。但这种传教的姿态未免过于二元地将现代与传统对立了起来。

实业家卢作孚看到了中国乡村在面临工业化竞争之时的小农经济的崩溃，他认为解决乡村问题应从振兴产业出发，通过发展乡村的生产事业和基础设施的建设，从而"赶快将这一个乡村现代化起来"。卢作孚对于中国乡村的社会结构也有深刻的认识，在论述中国人的两重社会生活时，提出"人每责备中国人只知有家庭，不知有社会；实则中国人只有家庭，没有社会，家庭就是中国人的社会。"[2] 在此认识的基础上，卢作孚提出了建设现代集团生活的理念，希望以社会为核心的新型乡村社会结构能替代以家庭为核心的传统乡村的社会结构。二十世纪二十年代末始，卢作孚主持和推动了以重庆北碚为中心的嘉陵江三峡地区三十多个乡镇的乡村建设运动。在北碚这个中古时代的乡镇，他首先建设新的温泉公园，让民众感受美好和谐皆可亲手创造。他还创办正规学校，建图书馆、博物馆、运动场、医院、报社等，将"嘉陵江三峡布置成一个生产的区域，文化的区域，游览的区域。"[3] 他在北碚的乡村建设当中引入大量的生产事业，进而创造文化事业和社会公共事业，希望能将物质文明和精神文明带到传统乡村。

社会学家费孝通将大量的社会学性质的田野调查用于中国乡村的研究。在其早期的著作《乡土中国》中，他对于乡下人为人所诟病的"愚"和"私"做了独到的剖析，深入挖掘了中国社会的乡土本色。以"愚"而言，他认为乡下人的"愚"实则是乡下人在文化人常用的文字体系之外另外建构了一套更为直接的、更倾向于会意的"特殊语言"体系。以"私"而言，他以石子入水所产生的涟漪作比，认为中国社会是以"己"为中心根据远近亲疏关系向外推的差序格局。在其著作《江村经济》中，费孝通进一步从乡村经济层面对传统乡村的社会结构做出分析。他选取了太湖边吴江开弦弓村作为其研究标本，详细描述了这一处在变迁过程中的典型传统乡村的消费、生产、分配、交换的经济体系，以及这一经济体系与特定地理环境、社会结构之间的关系。这些田野调查让他深刻认识到"中国农村真正的问题是人民饥饿的问题"。[4]

used by the more educated people. As for "selfishness", he applied the analogy of the ripples created by a stone dropped into water, and suggested that Chinese society reflected a pattern of different sequence centered on the "self" and pushed outward according to proximity and affinity. In his another book "Peasant Life in China - A Field Study of Country Life in the Yangtze Valley", he took a further step to make an analysis of traditional rural social structure from the perspective of rural economy. He chose the Kaixiangong village beside the Tai Lake as his study sample, he elaborated the economic system of consumption, production, distribution, and exchange in this typical traditional village in the process of change, and the relationship between this economic system and the specific geographical environment and social structure. These field investigations made him profoundly aware of that "it is the hunger of the people that is the real issue in China." *(Fei Xiaotong. Peasant Life in China - A Field Study of Country Life in the Yangtze Valley. 2001. P236)*

The most influential scholar in the study of the rural issues in the period of the Republic of China must be Liang Shumin. In the May Fourth New Culture Movement, he dared to stand up for Confucianism amid the call to overthrow it. As a pioneer of Neo-Confucianism, he was committed to seeking ways from traditional Chinese culture to transform old China. Rural construction is a realistic way for him to realize China's social transformation. From 1931 until the beginning of the War of Resistance against Japanese Agression, Liang Shumin had been conducting rural construction experiments in Zouping, Shandong province. His book "Rural Construction Theory" was published in 1937. According to Liang Shumin, "China's problem is nothing but truly a cultural disorder - an extremely serious cultural disorder." *(The Complete Works of Liang Shuming. 2005. P164)* What is culture? "The culture of a society is based on the organizational structure of the society, and the legal system and etiquette are the most important part of it." *(The Complete Works of Liang Shuming. 2005. P162)* In an era when China's old social structure, which had rarely changed over the centuries, was about to collapse due to serious cultural disorders, Liang Shumin realized that the countryside was the root of Chinese culture, "China's inherent society is an ethic society and a related society; There is still a little ethos and atmosphere left in the countryside, unlike it was in the city, which has been destroyed totally! ...so called 'as rituals lost, seek in the country', there still retains many inherent customs in the countryside." *(The Complete Works of Liang Shuming. 2005. P316)* As an organization that tied up most of the Chinese people, the village, in Liang Shumin's view, was the most suitable structure to start establishing Chinese new social organization. This new organization would come from the supplementary transformation of the tradition of the countryside by spontaneous and autonomous "village covenant". At the same time, he proposed to replace the administrative power with educational power by the establishment of the peasant schools to implement a "Theocracy" like system. The peasant schools were also considered to be the embryonic form of future rural autonomous organizations. For Liang Shumin, the rural construction experiment was in fact his exploration of Chinese social construction. "So rural construction is not the construction of villages, but the construction of the whole Chinese society, or a kind of nation-foundation movement." *(The Complete Works of Liang Shuming. 2005. P161)*

民国时期在乡村问题研究方面最有影响力的应该是梁漱溟先生了。梁漱溟在"五四"新文化运动的一片打倒孔家店的呼声中敢于公开站出来维护儒家学说，并以新儒家开山者的姿态，致力于从中国传统文化中寻求改造旧中国的方法，乡村建设正是他实现中国社会改造的一条现实途径。梁漱溟于 1931 年开始直到抗战开始一直在山东邹平进行乡村建设实验，并在 1937 年出版《乡村建设理论》。梁漱溟认为"中国问题并不是什么旁的问题，就是文化失调 —— 极严重的文化失调。"[5] 何为文化？"社会之文化要以其社会之组织构造为骨干，而法制、礼俗实居文化之最重要部分。"[6] 在中国的千百年鲜有变化的旧社会构造因严重的文化失调而面临崩溃的时代，梁漱溟看到乡村是中国文化的根，"中国固有的社会是一种伦理的社会、情谊的社会；这种风气、这种意味，在乡村还有一点，不像都市中已被摧残无余！……所谓'礼失而求诸野'，在乡村中还保留着许多固有风气。"[7] 乡村作为羁縻大部分中国人民的组织，在梁漱溟看来是最适合入手建立中国社会组织的新构造。而这个新组织即来自于对乡村传统的自发和自治的"乡约"的补充改造。同时建设乡农学校以教育力量代替行政力量，实行"政教合一"，当然乡农学校也是未来乡村自治组织的雏形。对于梁漱溟而言，乡村建设实验实则是他对中国社会建设的探索，"所以乡村建设，实非建设乡村，而意在整个中国社会之建设，或可云一种建国运动。"[8]

[1] 梁漱溟 . 乡村建设理论 [M]// 中国文化书院学术委员会编 . 梁漱溟全集 : 第二卷 . 第 2 版 . 济南 : 山东人民出版社 , 2005: 153.
[2] 凌耀伦，熊甫编 . 卢作孚文集 [M]. 北京 : 北京大学出版社 , 1999: 323.
[3] 同 [2]: 343.
[4] 费孝通 . 江村经济 [M]. 北京 : 商务印书馆 , 2001: 236.
[5] 同 [1]: 164.
[6] 同 [1]: 162.
[7] 同 [1]: 316.
[8] 同 [1]: 161.

Country & City

—— A Road to Urbanization in China

The study and practice of the countryside during the period of Republic of China embodies the close connection between rural issues and Chinese society. The problems of China at that time were almost equivalent to rural problems. Whether in the city or country, "Chinese society is fundamentally rural". *(Fei Xiaotong. From the Soil: The Foundations of Chinese Society, & The institutions for Reproduction. 1998. P6)* Despite the political and economic aspects, at least in terms of the structural characteristics of the living space, the spatial structure of ancient Chinese towns and cities was almost indistinguishable from that of rural settlements, if the heavy defense walls were taken out of the picture. According to the traditional stratification, the social ranking of a craftsman or a merchant in the cities was even lower than a peasant who possessed land. And the upper class of urban residents was generally the landowners who owned large pieces of land in the countryside. The enormous inertia of the Chinese social structure, which had lasted for thousands of years, resisted tenaciously the gradual imported influences from the West after the Opium War on both political and cultural front. So even during the period of Republic, it was fair to say that the vast land of China was still endless countryside, which was at least not an exaggeration on a psychological level.

Although there had been already some industrialization development in some major cities in the Republic of China, especially in coastal cities like Shanghai which had been opened earlier. According to the National Bureau of Statistics, the urbanization rate in the early years after the founding of the People's Republic of China was only 10%. In large cities like Beijing or Nanjing

乡村与城市

—— 中国城镇化之路

　　民国时期的乡村研究和实践，体现了乡村问题和中国社会的密切勾连，当时的中国的问题几乎就等同于乡村问题。不管在城市还是乡村，"从基层上看去，中国社会是乡土性的。"[1]抛开那些政治经济方面的问题，起码就居住空间的结构特征来看，如果去掉那厚重的城墙，中国古代城镇的空间结构和乡村群落几乎没有本质的区别。而从传统的士农工商的阶级划分来看，城市当中的手工艺者或商人的地位也都不如手中有土地的农民，或者说城市当中的上层居民一般来说都是在乡村仍有着大量土地的地主。中国沿袭了几千年的社会结构所具有的巨大惯性顽强地抵抗着鸦片战争之后从西方逐渐输入的东西，不管在政治层面还是文化层面。即使进入民国，可以说偌大的中国土地上仍是绵延无际的乡村，至少在心理层面这么说是并无夸张的。

　　尽管在这一片乡村大地上，民国时期的一些大城市已经颇有一些工业化的发展，尤其像上海这样较早开埠的沿海城市。但是以国家统计局的数据来看，解放初年的城市化率不过 10%。像北京或南京这样的大城市，甚至旧时的城墙尚在，很难说城市对于整体社会的各方面影响在跨过老城墙这个物理边界之后还能剩多少。没有广泛的社会参与，缺乏强势的政府，零星的工业化并不能对城市的重新塑造起到太大作用，也远不足以对全国范围内的乡村面貌产生比较根本的影响。所以民国的各种

where the ancient city walls still existed, it was difficult to judge how much of the urbanization influence remained beyond the physical boundary of the old city wall. Without broad social participation and a powerful government, scattered industrialization was not capable of reshaping the city, nor was it sufficient to have a more fundamental impact on the rural landscape nationwide. Therefore, the various rural experiments conducted during the period of Republic of China were all about finding ways through political regime, industrial development and civilian education, while these strategies were really indistinguishable between urban and rural areas at a time when urbanization was not prominent.

The social organization of the countryside changed dramatically after the founding of the People's Republic of China with the successive major policies of land reform and collectivization, which was a very complex development process closely related to the political and economic changes of that time. In terms of land, the land reform policies of the early years maximized equalization of rural land rights that allowed more poor peasants or employed peasants to acquire land. With the loss of land, the gentry class lost the clan power which had been attached to it. And the peasants who had just acquired land, immediately encountered a wave of agricultural cooperativization, which involved the transfer of land ownership to the collective and distribution of labor remuneration according to labor work points. The peasants' attachment to the countryside was essentially an attachment to the land. Once the transfer of land ownership from the individual peasant to the collective or commune, it was difficult to expect the peasants to establish that special emotional connection with the land.

As the tie with the land was weakened, the peasants' right to move was almost simultaneously restricted. The first Hukou registration law "Temporary Regulations on Urban Hukou Registration Management" enacted in 1951 actually guaranteed the freedom of people's movement, which was clearly stated in its first article. However, under the guidance of the development model centered on heavy industry, the first five-year plan of the new China began to implement the policy of unified purchase and sale of agricultural products. The "unified purchase" was a planned purchase of grain from peasants who were no longer allowed to buy and sell by themselves, while the "unified marketing" was a planned supply of grains to the urban residents. The implementation of this policy required a reliable control over the population data. The proportion of urban and rural population also needs to be fixed in order to prevent the plan from deviating from its predetermined goals. In 1958, the NPC Standing Committee signed the "Regulations of the People's Republic of China on Hukou Registration", which required peasants to apply to city's Hukou registration office and pass the examination to obtain the moving permission before doing so. The population migration was thus completely controlled by this measure. Since then, the rural policies that restricted the flow of rural population but did not provide social security of any sorts had led to a sharp decline of rural living conditions. The urban and rural populations began to form a dichotomous structure with a huge difference in status, i.e. city people who lived on the assigned food in the planned system and country people who grew grains to feed the former. Country people seemed to be labeled as inferior, and being a city people became the biggest dream of a country

乡村实验不外乎是从政治体制、产业兴邦、平民教育上去寻找出路，而这些策略在城市化并无彰显的时代实在是不分城市和乡村的。

乡村的社会组织关系在建国之后随着相继而来的土改和集体化等重大政策而发生了巨大的改变，这是一个非常复杂的、和当时的政治经济变革紧密相关的发展过程。就土地而言，建国初期的土改政策最大限度地平均了农村的地权，让更多的贫农和雇农获得了土地。随着土地的丧失，乡绅阶层失去了依附于此的宗族权力。但是刚获得土地没多久的农民马上迎来了农业合作化的浪潮，农业合作化是将土地所有权转给集体，按照劳动工分来分配劳动报酬。农民对于乡村的依附本质上是对于土地的依附，当土地权属从农民个体转移到集体或公社之后，很难再指望让农民与土地之间建立那种特殊的情感联系。

当与土地的关系减弱之后，农民的迁移权却几乎同时被限制了。建国初期于1951年所颁布的第一个全国性户籍法规《城市户口管理暂行条例》是保障人民迁徙自由的，这一点在条例的第一条即有说明。但是在以重工业为中心的发展模式主导下，新中国的第一个五年计划开始对农产品实行统购统销政策。"统购"是按计划向农民统一收购粮食，农民不再被允许私自进行粮食买卖，而"统销"是按计划向城市居民供应粮食。这一政策的实施，需要对于人口数据有可靠的掌控，甚至城乡人口的比例也需要固定下来，以保障计划不会偏离预设目标。而最为有效的措施就是严控迁徙的户籍管理制度。1958年全国人大常委会签署了《中华人民共和国户口登记条例》，该条例要求农民迁入城市前先向拟迁入城市的户口登记机关申请并在审查合格后才准予迁入，这也就彻底控制了人口迁徙。此后，束缚农村人口流动但又不提供生活保障的乡村政策使得乡村生活条件急剧下降。城市和乡村人口在身份认知上开始形成一种地位相差悬殊的二元结构，即体制内吃皇粮的城里人和种粮食供养城里人的乡下人。乡下人似乎被贴上了低人一等的标签，做一个城里人成为乡下人的最大梦想。在集体化到统购统销以及之后相关的乡村政策的推动下，传统乡村的社会组织关系从根本上瓦解了，只是这种瓦解的状态仍被户籍制度给包裹着以不至于散架。这是一个乡村信仰被迅速侵蚀的阶段。

从二十世纪七十年代末的改革开放开始，中国的工业化和现代化发展开始逐渐加速。虽然户籍制度在二十世纪八九十年代还是严格限制着城市人口的发展，但是已经允许人员自由流动，大量的乡村人口像候鸟一样定期来到城市打工，农民工一

people. Collectivization, unified purchasing and marketing, and the various policies related to the countryside afterwards, actually fundamentally dismantled the social organization of the traditional countryside, the state of which was only supported by the Hukou registration system to keep it from falling apart. This was a period when the beliefs of the countryside were completely devastated.

Starting from the reform and opening up in the late 1970s, China's industrialization and modernization began to accelerate gradually. Although the Hukou registration system still restricted the development of the urban population in the 1980s and 1990s, it started to allow the free movement of people. A large number of people from the countryside came to the cities regularly to work like migratory birds, which bred the term "migrant workers". In the new century, especially after joining the WTO, China became a synonym of global factory. The development of urbanization required the expropriation of a large amount of rural land for urban construction. Peasants who had lost their land in the process of land expropriation were naturally transformed into urban residents. The Hukou registration system gradually ceased to be a barrier to migration, and more peasants were attracted by the various opportunities in the cities and settled down. This is a time of great urbanization. According to the data, under the strict control of the Hukou registration system, the urbanization rate slowly increased by 10 percentage points within the thirty years after the founding of the People's Republic of China; it increased by another 10 percentage points within the twenty years from the reform and opening up till the end of the twentieth century; and it increased by 30 percentage points in the next twenty years of urbanization, showing an obvious acceleration. By 2020, China's urbanization rate has exceeded 60%.

The Chinese countryside has rapidly changed in the context of this unprecedented urbanization. The rural issue has been converted to be a problem closely related to urbanization. The success of urbanization and the decline of the countryside are like a pair of Siamese twins that cannot be separated. The biggest crisis brought by the success of urbanization to the countryside is the loss of rural population. With all the conveniences of city life that one cannot refuse, the countryside is basically unable to retain any of its young and strong population. Going to a village, a visitor would most likely to see the elderly, women and children. The issue of the elderly and children left behind has become one of the major rural problems of contemporary China. The burden of farming work has to be put on the shoulders of elderly in many places, which makes it common to see villagers working in the fields in their 70s and 80s. It is hard to imagine where these villages are headed after this generation fades out. In order to solve the rural poverty problem or improve the rural public service system such as centralizing and optimizing the allocation of educational and medical resources, the government has taken various measures to relocate villages and build new towns, such as moving villagers out from inconvenient areas to newly built centralized settlement, and merging small natural villages into one large village. These measures are certainly welcomed by villagers. But those remote villages usually keeping more traditional inheritance have to face the destiny of complete abandonment. Often times a village appearing to be beautiful layers of earthen

词也应时而生。进入新世纪，尤其是加入世贸组织之后，中国成了全球工厂的代名词。城镇化的发展需要征用大量的乡村用地将之转变为城市建设用地，在征地过程中失去土地的农民自然就转变为城市居民。户籍制度也逐渐不再成为人口迁徙的障碍，更多的农民被城市里的各种机会所吸引而定居下来不再返乡。这是一个城镇化大发展的时代。从数据来看，在严格的户籍制度管控之下，城镇化率在建国后的三十年间缓慢地增加了 10 个百分点；改革开放到二十世纪末的二十年间城镇化率又提高了 10 个百分点；而接下来的城镇化大发展的二十年里，城镇化率则提高了 30 个百分点，呈现出非常明显的加速态势。截至 2020 年，中国的城镇化率已经超过 60%。

　　中国乡村在这史无前例的城镇化的背景下快速异变。乡村问题开始转变为一个和城镇化密切相关的问题，城镇化的成功和乡村的衰败就像一对难以分割的连体婴。城镇化的成功给乡村带来的最大危机是乡村人口的流失。城市生活的各种便利让人不能拒绝，乡村基本上不能留住任何青壮年人口。去到几乎任何乡村，能看到的村民多是老幼妇孺，留守老人和留守儿童的问题已经成为当代的主要乡村问题之一。村子里的农活几乎只能由老人们来完成，七八十岁的老人还在地里干活的比比皆是，很难想象在这一代人凋零之后，这些村庄会走向何方。出于解决乡村贫困问题或是改善乡村公共服务体系的目的，譬如集中优化配置教育资源或是医疗资源，政府也有各种迁村建镇的行动，将交通不便地区的村民迁移出来集中安置，或是将一些小的自然村合并为大的行政村。这样的行动当然受到许多村民的欢迎，但是往往偏僻一些的村子还能有一些原汁原味的传统面貌，现在却不得不面对整体废弃的命运。远看是层层叠叠的瓦屋泥墙，走近却是残垣断壁寥无人烟，当看到这些村子正在被时光慢慢销蚀而成为如历史废墟一般，就会明白所有的乡村保护策略在没有人口的情况下，都将成为空谈。城镇化的进程不仅虹吸乡村人口，还有乡村的各种自然资源。在新世纪的前十年，为了自然资源的开采而破坏乡村环境的案例曾经屡见不鲜，甚至山村里的古树也是城市觊觎的对象。城市建设过程当中，为了创建生态优美的城市环境，最方便快速的就是将那些深山老林里的大树搬进城里。曾经荫蔽了多少代顽童的村口大树就这样一夜之间被不法分子转卖到了城市，而在这背井离乡的过程中也不知道有多少大树能存活下来。

　　乡村在城镇化进程当中不断向城市提供最宝贵的人口和资源，反过来也在不断地接受城市的价值观。伴随着城镇化的进程，乡村的面貌与以前相比大异其趣。城市的规划被照搬到乡村，新的农居房往往被规划成行列式的一排排完全复制的小别

walls and tile roofs when looking from distance, turned out to be ruins without any dwellers when approached. Witnessing these villages being slowly eroded by time into historical ruins, it is evident that all rural preservation strategies will become empty talk without population. The process of urbanization not only siphons off the rural population, but also the natural resources of the countryside. In the first decade of the new century, cases of destroying the rural environment for the sake of exploitation of natural resources were common, and even the old trees in the mountain villages were coveted by the cities. In the process of urban construction, in order to create an ecologically beautiful urban environment, the most convenient and fast way is to move those trees deep in the forest into the city. The old trees at the entrance of the village which had peacefully sheltered many generations were thus smuggled to cities overnight. Sadly no one knows how many trees could survive this abduction.

The countryside provides the city the most valuable like population and resources in the process of urbanization, and in turn embraces the value of the city. The look of the countryside has changed dramatically. Urban planning has been copied and pasted to the countryside, with new rural houses laid out in grids, which is a symbol of being affluent. In villages where the government is less involved, country people also did their best to rebuild their old houses with their hard-earned income from working in the city. The house is the top priority in the countryside. Whether or not a country people is successful making a living outside is judged by the house back at home. That seems to be their fate. A country people got used to the city's living standard when working and living there, and got brainwashed by the city's aesthetic interest as well. By the city standard, their old houses look shabby and low. So the compound handed down from the ancestors were knocked down, while their new houses got taller and taller, three stories, four stories, and five-stories. The craft of building traditional houses is of little use in the construction of these foreign styled houses. What the migrant workers have learned in the city is to bend steel bars and make mould. Houses built by the reinforced concrete are firm, fast built and economical. In a few short decades, the image of traditional village of earthen walls and black tiles has been quickly replaced by that of houses built by cheap industrial materials that country people could easily obtain in the rural market. With the help of rapid dissemination of ideas and efficiency of construction equipment, the contemporary urbanization process caused a devastating subversion of rural traditional image.

As observing China's rural issues, from the exploration of China's development path through rural experiments in the period of Republic of China, to the destruction of rural beliefs by the establishment of the urban-rural dualistic society in the early years of the People's Republic of China, then to the decay of traditional villages in the rapid urbanization process after the reform and opening up, the focus on rural issues has shifted largely from the early socio-political perspective to the contemporary aesthetic perspective.

墅，这是富裕乡村的象征。在政府介入较少的乡村，乡下人同样以自己在城里辛苦打工的收入努力改造自家的老房。在乡村，房子是头等大事，外面打工是否有出息，就看家里的老房能否盖过同村人的风头，这似乎是乡下人的宿命。乡下人在城市打工和生活的日子里，学到了城市的生活标准，却也被城市的审美意趣所洗脑。以城市的标准来看，自家的老房显得又破又矮。所以祖上传下来的合院被推倒，三层 —— 四层 —— 五层，自建的农民房越来越高。以前的造房手艺在这些洋房的建造上也派不上什么用场了，而且农民工们在城里建房子学到的是弯钢筋和搭模板，钢筋混凝土的房子又牢又快又省事。短短几十年，泥墙青瓦的传统村落意象迅速地被农民们用他们在农村市场上能取得的廉价工业建材所拼搭而成的自建房取代。当代的城市化进程倚赖于思想传播之疾与建设器具之利，对于乡村传统意象的颠覆实在是摧枯拉朽。

纵观中国的乡村问题，从民国时期通过乡村实验来探寻中国的发展道路，到中华人民共和国成立初期的城乡二元社会的建立对于乡村信仰的摧毁，再到改革开放之后的快速城镇化过程中的传统乡村的衰败，乡村问题从早期的社会政治的角度更多地转向当代的美学角度。

[1] 费孝通 . 乡土中国 生育制度 [M]. 北京 : 北京大学出版社 , 1998: 6.

this enunciation to enrich the overall image structure.

But when we enunciate a rural image, such as a sentence like "a village surrounded by a stream behind a stone bridge", the picture generated in the mind of a Chinese or an English may be completely different. In an English mind, it may be a Picturesque scene of thick verdant foliage behind the stone bridge and luxuriant meadow by the stream; while in a Chinese mind, it may be a picture scroll of a secluded mountain village where peasants' wives doing laundry on stone slabs in the stream and ducks floating down through the moon arch of the stone bridge with misty mountains as background. For the same idyllic enunciation, the English countryside exudes a romantic and quiet aristocratic vibe, while the Chinese countryside reflects a simple and reclusive demeanor of high scholar. So it is obvious there is a deep structure that determines people's aesthetic perspective towards the countryside beneath the surface description of a rural image. In linguistics, surface structure is the form of a sentence, and deep structure is the implication. As a kind of psychological cognition, deep structure is the cognitive content, while surface structure is the concrete expression of the cognitive content. Borrowing from the concept of "collective unconscious" of psychologist Jung, Levi-Strauss believes that all kinds of superficial cultural phenomena are products of mental structure of human kind, which is so called "deep structure of mind ".

In the process of struggling upward to the goal of "modernity", the surface structure has gradually become fragmented. The fissures and layers of fragments make the surface structure more and more difficult to penetrate, leaving the deep structure easily forgotten in contemporary times. For the Chinese rural image, it is like a river which was once clear enough to see the riverbed even with aquatic plants; but the contemporary urbanization in China is like a flood that has rushed mud, sand, stones and debris from everywhere into the river, making it impossible to see the riverbed - that is, the deep structure of the countryside. Although a flood cannot change the riverbed, it does make it difficult to observe. Structuralists believe that structure has little connection with time, and that history is nothing but the change of structure over time. Such point of view may be too static, but it provides us a chance to retrace the aesthetic image of the Chinese countryside from a historical perspective. In comparing the deep structure of Chinese and Western cultures, Sun Longji takes "dynamic" and "static" as a comparison: "The 'deep structure' of Western culture has a dynamic 'purpose' intention, i.e., a will to unlimited power, therefore any 'change' leads to constant transcendence and progress...As for the 'deep structure' of Chinese culture, it has a static 'purpose' intention...that is to maintain the stability and invariability of the whole structure. So there can be changes in the 'surface structure' as much as possible, but any 'change' cannot lead to progress or transcendence." *(Sun Longji. The Deep Structure of Chinese Culture. 2004. P10)* Compared to the West, Chinese social and cultural structures have always been super-stable, which makes it not too inappropriate to use a structuralist approach to study the Chinese countryside.

Since the pre Qin Dynasty, Chinese traditional Confucianism and Taoism have focused

但是当我们在陈述同一个乡村意象的时候，譬如用"石桥后面的一个溪水环绕的村子"这样一句话来形容某个村庄的时候，在中国人和英国人心里生成的乡村意象却可能是完全不同的。在英国人心中，可能是石桥后面浓荫翳日，溪边芳草连天，一片苍翠欲滴的画意派的景象；而在中国人心中，可能却是透过石桥的月洞看到几个农妇在溪边的青石板上洗衣，溪中漂来一片麻鸭，而远山如黛，是一幅隐逸的山村图卷。同样是田园风光，英国的乡村田园给人一种浪漫恬静的贵族气息，中国的乡村给人一种质朴归隐的高士风度。很显然在乡村意象的表层描述下面有一个深层的结构决定着人们对于乡村的审美角度。在语言学当中表层结构是语句的形式，深层结构是语句的意义。作为一种心理认知，深层结构就是认识的内容，而表层结构是认识内容的具体表述。列维 - 斯特劳斯借用心理学家荣格（Jung）的"集体无意识"的概念，认为各种表层的文化现象都是人类心灵结构的产物，即所谓的"心智的深层结构"。

在一个向"现代性"的目标向上挣扎的过程当中，表层结构逐渐碎片化，碎片的裂隙和层叠使得表层结构越发难以被穿透，以至于深层结构在当代往往容易被人遗忘。对于中国的乡村意象来说，这就好比是一条河，曾经河水清澈，即使有一些水草也能看到河床；而当代中国的城镇化如同洪水过境一般，将各处的泥沙、石块、杂物都一股脑儿地冲入这条河流之中，让人看不清河床 —— 即乡村意象的那个深层结构。虽然洪水不能改变承托这条河流的河床，但是在洪水期观察这条河的河床的确是件难办的事情。结构主义者认为结构与时间无关，历史无非是结构在时间中的变化。这样的观点或许过于静止，但是给予我们从历史角度来回溯中国乡村的美学意象的可能性。孙隆基在对比中西方文化的深层结构时以"动静"为比较："西方文化的'深层结构'具有动态的'目的'意向性，亦即是一股趋向无限的权力意志，因此，任何'变动'都导致不断超越与不断进步……至于中国文化的'深层结构'，则具有静态的'目的'意向性……就是维持整个结构之平稳与不变。因此，在'表层结构'中尽可以出现变动，但是，任何'变动'总不能导致进步与超越。"[2] 和西方相比，中国的社会和文化结构一直有着超稳定的结构，这也使得用结构主义的方法来研究中国乡村并无太大的不妥。

先秦以来，中国传统的儒道两家就注重修身养性。道家通过"修身"以求心静无我，进而达到"天人合一"；儒家则认为"君子之守，修其身而天下平"，通过"修身"以求心定。孙隆基在研究中国文化的深层结构时指出，中国人是用"身"来指称自己的，

on self-cultivation. Taoists seek peace of mind and selflessness through "self-cultivation", and ultimately "the unity of man and nature"; Confucianists believe that "a gentleman shall be self-cultivating in order to make a peaceful world" (by Mencius), shall strive for being resolute in mind through "self-cultivation". When studying the deep structure of Chinese culture, Sun Longji pointed out that Chinese like to use "body" to refer to themselves. Compared with the Western concept of "personality", Chinese obviously have a strong "body" tendency: "the existence of so called 'physicality', refers to directing the intention of whole life toward satisfying the needs of the 'body'." *(Sun Longji. The Deep Structure of Chinese Culture. 2004. P25)* Characterized by the "physicality" rooted from the deep structure of Chinese culture, the deep structure of Chinese rural image first shows a soiled temperament. Chinese peasants' feelings for the land are more like an infant's subconscious generated from its attachment to mother's "body". That is a kind of collective unconsciousness starting from the "body" and leading to the "heart mind", that longs to be close to the land. Fei Xiaotong once said, "We often say that country people are figuratively as well as literally 'soiled'(tuqi). Although this label may seem disrespectful, the character meaning 'soil'(tu) is appropriately used here." *(Fei Xiaotong. From the Soil: The Foundations of Chinese Society, & The institutions for Reproduction. 1998. P6)* The so-called soiled cannot be simply understood as the opposite of "foreign style". It has a more profound meaning. It is with land that one can grow food or burning bricks and tiles to build houses. Being soiled was the only way for peasants to survive in the natural environment as well as maintain their physical health and spiritual pleasure.

Being soiled is the external expression of the inherent naturalness of Chinese people, but it was not all utilitarian. When illiterate peasants much embodied soiled quality, traditional literati that lived in the countryside as well actively participated in the process of perceiving and touching the land through farming on a daily basis. Nourished by the soil, traditional literati in turn sublimated the soiled into the Chinese literary spirit. From Zhuge Liang of the Three Kingdoms to Zeng Guofan of the late Qing Dynasty, farming-reading had been the guiding principles of Chinese literati. Based on farming-reading, soiled and literary temperament have coexisted without contradiction and built up the deep structure of aesthetic image of typical Chinese villages.

In China's agricultural society that has lasted for thousands of years, soil oriented is regarded as orthodox philosophy. Being soiled is a sense of righteousness, so the literary temperament rooted from being soiled determined that plainness was the primary element of rural aesthetic image. In this respect, the profound influence of the Song Dynasty's philosophy on Chinese culture was particularly significant. The Cheng Brothers and Zhu Xi of the Song Dynasty all advocated "knowing nature's law, exterminating human's desire", which was fiercely criticized as a feudal ethics though after the May Fourth Movement. But "in terms of aesthetics, it advocated the transcendence of aesthetics, which was to exclude the practical utilitarianism, so as to achieve the realm of aesthetics." *(Zheng Suhuai. History of Aesthetic Thoughts in Song Dynasty. 2007. P310)* Taking articles for example, Zhu Xi argued that "most of the ancient articles were plain in expression while the intentions were profound. The articles of the later

相对于西方的"人格"观念，中国人显然具有很强的"身体化"倾向："所谓'身体化'的存在，就是指将整个生活的意向都导向满足'身'之需要。"[3]基于中国文化的深层结构所显现的"身体化"特征，中国乡村意象的深层结构首先就显现出一种土气。中国农民对于土地的感情，更像是发自于一个婴儿对于母亲身体的依恋的那种潜意识，这是一种希望贴近土地的发于"身"而达于"心"的集体无意识。费孝通说"乡下人土气，虽则似乎带着几分藐视的意味，但这个土字却用得很好。"[4]所谓土气，不能简单地理解为"洋气"的反义，它有更为深刻的含义。有了土地之后，从种植口粮到烧砖制瓦建房子，土气是乡下人在自然环境中生存下来的唯一倚靠，并以此保持体格上的健康与思想道德上的愉悦。

土气是中国人内在自然性的外在表达，这也并不全是功利性的一面。如果说没有文化的农民身上更多体现的是土气，隐于乡野的传统文人则在日常身体力行地参与到对土地的感知和触摸的过程中，在土气滋养之上获得升华而形成中国的文气。从三国诸葛亮到晚清曾国藩，耕读持家是中国文人的修身准则。以耕读为本，土气和文气建构起典型中国乡村的审美意象的深层结构，并且并存不悖。

在中国延续几千年的农业社会中，以土为本是正道。土气也是正气，所以以土气为根本的文气决定了朴实是首要的乡村美学意象。在这方面，尤以宋代的理学思想对中国文化影响之深刻为甚。宋代的二程与朱熹都主张"存天理灭人欲"，虽然这个观点在"五四运动"之后被作为封建礼教而大加挞伐，但是"从美学的角度看，就是主张审美的超越性，就是要排除实用的功利性，从而达到审美的境界。"[5]譬如就文章而言，朱熹认为"古人文章，大率只是平说，而意自长。后人文章，务意多而酸涩。"[6]他批判黄庭坚的文章"一向求巧，反累正气"[7]，也就是说好的文章应该平实而不务巧。"但是文章的平实、闲澹并不是'柔弱'……并不是一味地提倡将平实变为柔弱之气。而是平实之中有阳刚之气，闲澹之中有华彩。那么，朱熹为什么提倡平实的文学风格呢？这与他的'文'与'道'的主张有关系……在朱熹看来，'文'是不可能'贯道'的，原因十分简单，就是'文'是'末'，而道是'本'。如果像韩愈所说'文者，贯道之器'的话就是本末倒置。那么，文与道是什么关系呢？朱熹认为应该是'文从道出'。"[8]

躬耕务农作为传统文人的生活方式的一部分，也意味着文气非常强调"身体"的在场。在中国传统山水画的艺术创作中，就一直强调由"身"的感知再到"心"

generations meant to be more meaningful instead appeared pompous and redundant." *(Zheng Suhuai. History of Aesthetic Thoughts in Song Dynasty. 2007. P321)* He criticized Huang Tingjian's articles as "always seeking clever words, but poor at righteousness", *(Zheng Suhuai. History of Aesthetic Thoughts in Song Dynasty. 2007. P321)* which meant that good articles should be plain as opposed to ornamental. "But the plainness and simplicity of an article is not 'softness'...not simply advocate turning plainness into softness. Rather there is masculinity in plainness, and Cadenza in simplicity. So, why did Zhu Xi advocate a plain literary style? This was related to his advocacy of 'literature' and 'Tao'...In Zhu Xi's view, 'literature' was impossible to 'hold the Tao'. The reason was quite simple. 'Literature' is the 'effect', and the Tao is the 'cause'. What Han Yu had argued, 'literature is a device holding the Tao', was regarded as having cause and effect reversed. Then, what is the relationship between literature and the Tao? Zhu Xi believed that 'literature shall be born of the 'Tao'." *(Zheng Suhuai. History of Aesthetic Thoughts in Song Dynasty. 2007. P323)*

Farming being part of the lifestyle of traditional literati, also means that literary temperament emphasizes the presence of the "body". The conversion from body perception to heart has always been the focal point in traditional Chinese landscape painting, which was exactly what the Tang Dynasty painter Zhang Zao meant, "learn from the nature and gain from the inherent comprehension". Liu Daochun of the Song Dynasty commented in his "Critics on the Famous Paintings of the Song Dynasty" that Fan Kuan "lived in the mountains and forests, often sitting still all day long and looking around for his interest." In his "Lofty Records of Forests and Streams", Guo Xi proposed "the body being in the mountains and streams to take what it feels". Therefore the primary requirement of Chinese landscape painting is to be physically present in the nature to take the most direct aesthetic reflection of the natural landscape. Only when the body has been imbued with the beauty of nature can the heart pursue a higher state of mind. As Guo Xi put in his famous method of "Three Types of Distance", "there are three types of distance observing the mountains: look from the bottom up to the top is called high distance; look from the front to the behind is called deep distance; look from the near to the distant is called level distance." Painters take deep reflection of nature through different physical views of looking up, looking down, and looking horizontally, guiding one's sight to the infinite space so as to distance the mind from the mundane world. Although "Three Types of Distance" is a method of landscape painting, it is the best depiction of the cosmology of Chinese literati who have ever been pursuing conversion from body to heart. There is similarity between self-cultivation in rural life and landscape painting. Body's farming work leads to the reconstruction of objective world's projection toward the traditional literati's mind, so as to make the heart joyful. Therefore, when we reexamine Tao Yuanming's poem "Picking chrysanthemums by the east fence, rests my gaze upon the south mountains", "picking chrysanthemums" sure is a physical act of presence, while "the south mountains" are not necessarily the objective world we see.

In summary, the temperament of traditional Chinese literati is oriented toward the soil and follows the laws of nature; while the writing seeks plainness and the painting seeks a distant

的转化，即唐代画家张璪所提"外师造化，中得心源"之意。宋代刘道醇在其《宋朝名画评》中评说范宽"居山林间，常危坐终日，纵目四顾，以求其趣。"郭熙则在《林泉高致》中提出"身即山川而取之"。可见山水画的首要要求是以身体亲临于自然环境中，对自然山水进行最直接的审美观照。在"身"得到自然之美的足够灌输之后，"心"才能去追求更高的意境。以郭熙著名的"三远"法观之："山有三远：自山下而仰山巅谓之高远；自山前而窥山后谓之深远；自近山而望远山谓之平远。"画家通过仰视、俯视、平视的不同身体观法，对自然进行深入的观照，将人的视线导向无限的空间从而在心灵上远离俗世。其"三远"之法虽是山水绘画之法，却可谓是对以身达心的中国文人宇宙观的一种最佳写照。传统的田园乡村生活的修身养性和山水画的创作有异曲同工之妙，田园的躬耕生活，引发客观世界对于文人心灵投射的重塑，从而获得心的愉悦。所以再看陶渊明的"采菊东篱下，悠然见南山"，"采菊"自然是身体的在场行为，"南山"却不一定是看到的客观世界。

于以上所述可见，中国传统文人气息以土为本，师法自然，文章以求平实之道，绘画以求心境之远。如文章，如绘画，散发着土气和文气的乡村栖居也是朴实而浪漫的。林语堂在其《吾国与吾民》中认为"乡村典型的生活，常被视为最理想的优美生活，农村的理想表现于艺术哲学及生活者，如此深植于一般中国人之意识中……中国生活典型之创始者能于原始的生活习惯与文明二者之间维持一平衡，其手段岂非巧妙？岂非此健全的本能，导使中国人崇尚农耕文明而厌恶机械技巧，并采取一种单纯的生活？岂非此健全的本能，发明人生的愉快而能使不致劳形役性，因而在绘画中，文学中，一代一代地宣扬着'归田'思想。"[9]林语堂在书中摘录郑板桥致其介弟之家书一段，颇能体现传统文人对于乡村田园的审美趣味实在是发于乡土之质朴，而达于旷远之心境。

吾弟所买宅，严紧密栗，处家最宜。唯是天井太小，见天不大，愚兄心思旷远，不乐居耳。是宅北至鹦鹉桥不过百步，鹦鹉桥至杏花楼不过三十步，其左右颇多隙地；幼时饮酒其旁，见一片荒地，半堤衰柳，断桥流水，破屋丛花，心窃乐之；若得制钱五十千，便可买地一大陂，他日结茅有在矣。吾意欲筑一土墙院子，门内多栽竹树草花，用碎砖铺曲径一条，以达二门；其内茅屋二间，一间坐客，一间作房，贮图书史籍、笔墨、砚瓦、酒钟、茶具其中，为良朋好友后生小子论文赋诗之所。其后住家主屋三间，厨房二间，奴子屋一间，共八间；俱用草苦，如此足矣。清晨日尚未出，望东海一片红霞；薄暮斜阳满树，立院中高处，便见烟水平桥。家中宴客，

state of mind. Like articles and paintings, the rural dwelling with a vibe of soiled and literacy is simple and romantic. In his book "My Country and My People", Lin Yutang argued that "as the rural mode of life was always regarded as the ideal. This rural ideal in art, philosophy and life, so deeply embedded in the Chinese general consciousness...Did the creators of the Chinese pattern of life do more wisely than they knew in maintaining a level between civilization and the primitive habits of living? Was it their sound instinct which guided them to choose the agricultural civilization, to hate mechanical ingenuity and love the simple ways of life, to invent the comforts of life without being enslaved by them, and to preach from generation to generation in their poetry, painting and literature the 'return to the farm'?" *(Lin Yutang. My Country and My People. 2006. P21)* In his book, Lin Yutang excerpted a paragraph from Zheng Banqiao's letter to his brother, which is a good example of the traditional literati's aesthetic taste toward the countryside, which was really based on the vernacular simplicity and reached a distant state of mind. *(Lin Yutang. My Country and My People. 2006. PP21-22)*

The house you bought is well-enclosed and indeed suitable for residence, only I feel the courtyard is too small, and when you look at the sky, it is not big enough. With my unfettered nature, I do not like it. Only a hundred steps north from this house, there is the Parrot Bridge, and another thirty steps from the Bridge is the Plum Tower, with vacant spaces all around. When I was drinking in this Tower in my young days, I used to look out and see the willow banks and the little wooden bridge with decrepit huts and wild flowers against a background of old city walls, and was quite fascinated by it. If you could get fifty thousand cash, you could buy a big lot for me to build my cottage there for my latter days. My intention is to build an earthen wall around it, and plant lots of bamboos and flowers and trees. I am going to have a garden path of paved pebbles leading from the gate to the house door. There will be two rooms, one for the parlour, one for the study, where I can keep books, paintings, brushes, ink-slabs, wine-kettle and tea service, and where I can discuss poetry and literature with some good friends and the younger generation. Behind this will be the family living-rooms, with three main rooms, two kitchens and one servants' room. Altogether there will be eight rooms, all covered with grass-sheds, and I shall be quite content. Early in the morning before sunrise, I could look east and see the red glow of the morning clouds, and at sunset, the sun will shine from behind the trees. When one stands upon a high place in the courtyard, one can already see the bridge and the clouds and waters in the distance, and when giving a party at night, one can see the lights of the neighbours outside the wall. This will be only thirty steps to your house on the south, and will be separated from the little garden on the east by a small creek. So it is quite ideal. Some may say, "This is indeed very comfortable, only there may be burglars." They do not know that burglars are also but poor people. I would open the door and invite them to come in, and discuss with them what they may share. Whatever there is, they can take away, and if nothing will really suit them, they may even take away the great Wang Hsienchih's old carpet to pawn it for a hundred cash. Please, my younger brother, bear this in mind, for this is your stupid brother's provision for spending a happy old age. I wonder whether I may have what I so desire.

墙外人亦望见灯火。南距汝家百三十步，东至小园仅一水，实为恒便。或曰："此等宅居甚适，只是怕盗贼。"不知盗贼亦穷民耳，开门延入，商量分惠，有甚么便拿甚么去；若一无所有，便王献之青毡亦可携取，质百钱救急也。吾弟留心此地，为狂兄娱老之资，不知可能遂愿否？[10]

[1] GUTTING G. French Philosophy in the Twentieth Century[M]. Cambridge: Cambridge University Press, 2001: 221.
[2] 孙隆基 . 中国文化的深层结构 [M]. 桂林：广西师范大学出版社 , 2004:10.
[3] 同 [2]: 25.
[4] 费孝通 . 乡土中国 生育制度 [M]. 北京：北京大学出版社 , 1998: 6.
[5] 郑苏淮 . 宋代美学思想史 [M]. 南昌：江西人民出版社 , 2007: 310.
[6] 同 [5]: 321.
[7] 同 [5]: 321.
[8] 同 [5]: 323.
[9] 林语堂 . 吾国与吾民 [M]. 西安：陕西师范大学出版社 , 2006: 21.
[10] 同 [9]: 21-22.

Archetype & Analogy

—— Study of the Rural Typology

For the study of the countryside, we can analyze the deep structure of the aesthetic image of the countryside purely from a social or cultural perspective; we can also conduct a kind of archaeological analysis on the various representations of the countryside. This archaeological activity attempts to empty the various historical dimensions attached to the representations and restore them to the types of abstract concepts. That is the main work of structural typology.

Italian architectural theorist Rossi extended the concept of typology from architecture to the urban field, incorporating the urban into a timeless concept of structure through typology. In his book "The Architecture of the City", he elaborated the issue of typology in depth. Typology was first introduced into the field of architecture by the French scholar Quatremère de Quincy, who explained the characteristics of typology by comparing "type" and "model", "The word 'type' represents not so much the image of a thing to be copied or perfectly imitated as the idea of an element that must itself serve as a rule for the model...The model, understood in terms of the practical execution of art, is an object that must be repeated such as it is; type, on the contrary, is an object according to which one can conceive works that do not resemble one another at all. Everything is precise and given in the model; everything is more or less vague in the type. Thus we see that the imitation of types involves nothing that feelings or spirit cannot recognize..." *(Aldo Rossi. The Architecture of the City. 1997. P40)* Based on Quincy's comparison, Rossi "defines the concept of type as something that is permanent and complex, a logical principle that is prior to form and that constitutes it." *(Aldo Rossi. The Architecture*

原型与类推

—— 乡村类型学研究

对于乡村，我们可以纯粹地从社会或文化的角度来分析乡村审美意象的深层结构；我们也可以对乡村的各种表象进行一种历史考古，这种考古活动试图抽空附着于表象上的各种历史维度，将表象还原到抽象概念的类型，这是结构类型学的主要工作。

意大利建筑理论家罗西（Rossi）将类型学的概念从建筑扩展到城市，通过类型学将城市纳入一种永恒的结构概念之中。他在其著作《城市建筑学》中阐述了类型学的问题。类型在建筑学领域最早由法国学者昆西（Quincy）提出，他通过对"类型"和"模型"的比较来说明类型的特征："'类型'这个词不能够像被用于模型规则的元素那样的概念来代表一件事物的可以被复制或完美模仿的形象……从艺术实际操作的角度来看，模型是可以被自我重复的；而类型则正好相反，根据类型可以去构想出完全不同的作品。模型中的一切是精确和给定的，而类型中的一切却多少是模糊的。因此我们看到，对类型的模仿不可能没有感觉和精神的参与……"[1] 罗西基于昆西的比较，将"类型概念定义为某种永恒和复杂的事物，一种先于形式且构成形式的逻辑原则。"[2] 就这个定义而言，西特（Sitte）在研究欧洲中世纪城市广场时给出了很好的例子，在其著作《城市规划：根据艺术性原则》的目录中可见一斑：

of the City. 1997. P40) In terms of this definition, Sitte gave good examples in his study of the urban squares of the medieval Europe, which was clearly shown in the index of his book "City Planning, According to Artistic Principles":

If these chapter titles joined together, isn't it about the principles of the formation of the type of European medieval squares?

Influenced by Sitte, European architectural scholars studied squares and streets in greater depth after World War II, returning to history to seek archetypes of these urban spaces in the trend of modernism. In his book "Town and Square", Zucker illustrated how different historical periods had characterized appropriate forms of square space for different functions, based on five square archetypes. Swiss architectural scholar Rob Krier took streets and squares as the main elements that constitute urban space, arguing that "knowing all the conceivable urban space typologies and the variety of possible facade designs in public spaces are further necessary prerequisites". *(Rob Krier. Town Space. 1988. P16)* Therefore, in his book "Stadtraum", he analyzed in great detail the spatial composition of Western towns and cities in terms of spatial plans, architectural sections and elevations in a way similar to the diagrammatic system used by the nineteenth-century French architectural scholar Durand in his study of architectural typology. Taking the plan as an example, from basic geometric archetypes like square, circle, and triangle, almost countless plan types of urban space can be derived through changes of different variables, such as transformation (rotation, cutting, addition, subtraction, merging, overlapping, distortion), regularity (regular or irregular), or enclosure mode (enclosed or open).

As far as type study concerned, type is a logical principle that precedes form and is abstract, but the absence of image representation of type also means that if type is used as a starting point of design, there is no formal language to refer to and one tends to fall into a kind of speculation of mysticism. So we see architectural scholars mentioned above resorting to the concept of archetypes used by Jung in his psychological studies. Archetype is derived from the Greek word meaning "first type" or "beginning type" and is also known as the primordial image, which is the basis for subsequent changes or combinations of forms. Jung believed that archetype was a constantly recurring image in mythology, religion, dreams, fantasy, and literature, which had originated from the long-term accumulation of collective psychological experiences of human generations and was the projection of history in the memory of the

如果把这些章节的题目联在一起，不就是关于欧洲中世纪广场这个类型的形成原则吗？

受到西特的影响，二战之后的欧洲建筑学者对于广场和街道做了更深入的研究，在现代主义的潮流中返回历史当中寻求这些城市空间的原型。朱克（Zucker）在其著作《城市与广场》中，在五种广场原型的基础上，阐述了不同历史时期如何因不同的功能而形成合宜的广场空间形式。瑞士建筑学者罗伯·克里尔（Rob Krier）将街道和广场看作是构成城市空间的主要元素，他认为"了解所有能想到的城市空间的类型和公共空间的各种可能的立面设计，是必要的先决条件"。[3]所以他在其著作《城市空间》一书中，以类似十九世纪法国建筑学者杜朗（Durand）研究建筑类型学时所采用的图示体系的方式，巨细靡遗地从空间平面、建筑剖面、建筑立面对西方城镇的空间构成进行了详尽的分析。以平面为例，从方、圆、三角这样的基本几何原型出发，通过不同变量的改造，如变形（旋转、切割、加减、合并、重叠、扭曲），规则度（规则或不规则），围合方式（围合或开放），几乎可以得出无数种城市空间的平面类型。

就类型研究而言，类型是一种先于形式的、抽象的逻辑原则，但是类型缺失表象也意味着如果以类型作为设计的起点，没有任何可以参照的形式语言而陷入一种神秘主义的猜测之中。所以我们看到上述建筑学者借助了荣格在心理学研究中使用的原型概念。原型源自希腊语，意思是"第一型"或"开始型"，也被称为原始意象，它是之后形式变化或组合的基础。荣格认为原型是神话、宗教、梦境、幻想、文学中不断重复出现的意象，它来源于人类世代的集体心理经验的长期积累，是历史在族群记忆中的投射，它能唤起观察者潜意识中的原始经验，使其产生深刻、强烈、非理性的情绪反应。在任何城市或者乡村的历史意象背后，我们可以认为存在一组简单的原型的集合，这些原型可以以形式、功能、技术等方式呈现。原型的使用必

ethnic groups, which could evoke the primitive experience from an observer's subconscious and cause a profound, intense, and irrational emotional response. Behind the historical image of any city or village, we may consider the existence of a collection of simple archetypes that can be presented in form, function, technology, etc. The use of archetypes must accomplish three goals in the design process: first, to provide a focused overview of the archetype; second, to describe the expressive potential inherent in the archetype; and third, to select an appropriate form for the archetype to maximize its expressive potential. It is through the aggregation of archetypes of urban space such as squares and streets, certainly including various architectural archetypes like churches, after the formal changes in different historical periods, did a European medieval town form a complete and unique urban artifact. As Rossi stated, "the beauty of the Gothic city appears precisely in that it is an extraordinary urban artifact whose uniqueness is clearly recognizable in its components. Through our investigation of the parts of this city we grasp its beauty: it too participates in a system. There is nothing more false than an organic or spontaneous definition of the Gothic city. " *(Aldo Rossi. The Architecture of the City. 1997. PP53-54)*

Most cities and villages are not the products of one generation or one period of time, but rather the people, who have lived there at each stage of history, have added things to them. They are "artifacts" with historical continuity. The historical continuity of European city has been broken in the modernism trend. A city, once a complete artifact, was ripped by lost spaces that need to be patched. In response to the functional-formal paradigm of modernism, Rossi described his design approach in his article "An Analogical Architecture" as a "logical-formal"conversion. He introduced Jung's concept of "analogy" to elaborate this approach, "'logical' thought is what is expressed in words directed to the outside world in the form of discourse. 'Analogical' thought I sensed yet unreal, imagined yet silently; it is not a discourse but rather a meditation on themes of the past, an interior monologue. Logical thought is 'thinking in words.' Analogical thought is archaic, unexpressed, and practically inexpressible in words." *(Aldo Rossi. An Analogical Architecture. //Kate Nesbitt. Theorizing a New Agenda for Architecture: An Anthology of Architectural Theory 1965-1995. 1996. P349)* Rossi felt in Jung's definition of "analogy" "a different sense of history, which was not simply composed of facts, but of a series of things, of emotional objects that had been remembered or used in design." This means that the charm of the past, when processed and embellished by human memory, can often surpass that of the present. This is why Halbwachs has argued that "There is a kind of retrospective mirage by which a great number of us persuade ourselves that the world of today has less color and is less interesting than it was in the past, in particular regarding our childhood and youth." *(Maurice Halbwachs. On Collective Memory. Edited and translated by Lewis Coser. 1992. P48)* Rossi borrowed the theory of "collective memory" from Halbwachs, and believed that "city itself is the collective memory of its people, and like memory it is associated with objects and places. The city is the locus of the collective memory." *(Aldo Rossi. The Architecture of the City. 1997. P130)* So Rossi said, "I am referring rather to familiar objects, whose form and position are already fixed, but whose meanings may be changed. Barns, stables, sheds, workshops, etc. Archetypal objects whose common emotional appeal reveals timeless

须在设计过程中完成三个目标：一是对原型进行集中的概述；二是对原型所内含的表现潜能进行描述；三是为原型选择合宜的形式以最大限度地发挥它的表现潜能。中世纪欧洲城市正是由广场和街道这样的城市空间的原型，当然也包括如教堂那样的各种建筑原型，在历经不同历史时期的形式变化之后，聚合成为一个完整而独特的建筑体的。如罗西所言，"哥特城市的美丽恰恰展现于它是一个非凡的城市建筑体，其独一性能清晰地在它的组成元素中辨析出来。通过对于城市各个部分的考察，我们感悟到它的美丽：即它也参与到系统之中。再也没有比将哥特城市定义为有机的或自发的更为错误的观点了。"[4]

　　绝大部分的城市或是村庄都不是一代人或一个时期的产物，而是每个历史阶段的居住于此的人们都为此增加过新的内容，这是具有历史连续性的一个"建筑体"。在现代主义大潮中，欧洲城市曾经的历史连续性断裂了，作为一个完整建筑体的城市出现了需要被填充的真空地带。针对现代主义的"功能 - 形式"的范式，罗西在他的《一种类推建筑》一文中将他的设计方法称为是一种"逻辑 - 形式"的转换操作，他引入了荣格关于"类推"的思维概念来阐述这个方法："'逻辑'思维是以话语的形式用语言表达以直接指向外部世界的。'类推'的思维是被感知的但并不真实，是被想象的但也是沉默的，它不是一种话语，而是一种对过去主题的沉思，一种内心独白。逻辑思维是'用语言思考'，类推思维是古老的、未表达的、实际上也是无法用语言表达的。"[5] 罗西从荣格的"类推"定义中感受到"一种不同的历史感，这种历史感并不简单地由事实构成，而是由一系列事物、由被记忆或在设计中使用的情感对象构成。"这就意味着经由人的记忆加工和美化之后，过去的魅力往往能超越当下。所以哈布瓦赫（Halbwachs）认为"社会自身总是让身处其中的个人产生一种幻象：似乎今天的世界和过去的世界相比，总有些莫名的不完满。"[6] 罗西借用哈布瓦赫的"集体记忆"的理论，认为"城市自身就是市民们的集体记忆，和记忆一样，城市也和物体与场所相关联，城市是集合记忆的场所。"[7] 所以罗西"情愿参照那些形式和位置已经固定的熟悉物体，尽管它们的意义可能会改变。谷仓、马厩、棚屋、工作坊，等等，这些原型物体的共同情感诉求揭示了不受时间影响的关注。"[8] 罗西的"类推城市"即是通过集体记忆堆积起来的各种城市纪念物、各种历史片段的集合，可以说类推城市即是一次立足现在面对过去的重构。

　　基于集体的心理经验，原型概念从城市的元素层面，类推概念从城市的整体层面，一起构成了结构类型学的主要设计模型，这是一种和历史有着深切关联的思考模型。

concerns." *(Aldo Rossi. An Analogical Architecture. //Kate Nesbitt. Theorizing a New Agenda for Architecture: An Anthology of Architectural Theory 1965-1995. 1996. P349)* Rossi's "analogous city" is thus a collection of various urban monuments and historical fragments accumulated through collective memory. It is fair to say that an analogous city is a reconstruction based on the present and facing the past.

Based on collective psychological experience, the concept of archetypes concerning the elemental level of the city, and the concept of analogy concerning the holistic level of the city, together constitute the main design model of structural typology, a model of thinking deeply connected to history. As far as typological research is concerned, cities and villages are not fundamentally different from each other. If conducting an archaeological study of the Chinese countryside by this model, the courtyard, the open hall, the farmland, the drying ground, the ancestral hall, the stage, the ancient tree, the stone bridge, etc., these types assembled into a rural artifact. Compared to the city, the rural life pattern has certain randomness, which makes the charm of the past presented by the rural types much lively as opposed to pathological. Here in the countryside there is no such isolated and "pathological" permanences stated by Rossi. Take the ancestral hall for example, it was once a place where all activities were carried out to serve the purpose of glorifying the house. So it was a place where the house rules were executed and where the sons of different branches were called upon to study. A troupe would be invited to the ancestral hall to perform during the New Year and other important festivals to show the prosperity of the clan. In today's society where patriarchal system has degenerated, the function of law enforcement and indoctrination within the clan no longer exists. Although an ancestral hall might still be happening, with the elderly left behind gathering and playing mahjong under the ancestral tablets. As ludicrous as it may look, the vitality of a rural type has been validated.

就类型学研究而言，城市和乡村没有本质的不同。借助这个模型对中国乡村进行考古，合院、明堂、农田、晒场、宗祠、戏台、古树、石桥等等，正是类型集合成为一个乡村建筑体。和城市相比，乡村生活模式具有某种随意性，这使得乡村的类型所呈现的那种过去的魅力并不是病态的，而是鲜活的。在这里，没有罗西所提的那种孤立的"病态的"经久要素。以宗祠为例，曾经这里一切的活动出于光耀门楣的目的，所以这里既是执行族规的地方，也是将族中子弟合在一起办私塾的地方，逢年过节还常会请戏班表演以彰显本族的兴旺。在宗族社会已经退化的今天，宗族内部执法和教化的功能不复存在了，但很多宗祠仍旧很热闹，进去会发现祖宗的牌位下是一桌桌的留守老人的麻将。虽然看上去有点啼笑皆非，但是很好地印证了乡村类型的生命力。

[1] ROSSI A. The Architecture of the City[M]. Boston: MIT Press, 1997: 40.
[2] 同 [1]: 40.
[3] KRIER R. Town Space[M]. New York: Rizzoli, 1988: 16.
[4] 同 [1]: 53-54.
[5] ROSSI A. An Analogical Architecture[M]. //NESBITT K. Theorizing a New Agenda for Architecture: An Anthology of architectural Theory 1965-1995. Princeton: Princeton Architectural Press, 1996: 349.
[6] HALBWACHS M. On Collective Memory[M]. Edited and translated by COSER L. Chicago: University of Chicago Press, 1992: 48.
[7] 同 [1]: 130.
[8] 同 [5]: 349.

Symbol & Allegory

—— Struggle in Modernity

In the current trend of urbanization, we also need to be aware of the fact that the changes of this era are more profound and disruptive than ever before. Western society has been reshaped since the Medieval Ages through successive ideological changes in Renaissance, Reformation, and Enlightenment. Even though it has been in the process of constant change, Western society still faces great confusion when encountered with the modernity spawned by the capitalist era. On the contrary, Chinese society had maintained a stable cultural structure until the Opium War, relying on the dominance of Confucianism, which had been constantly strengthened since the Tang and Song dynasties, and the systematic guarantee of the patriarchal society and the imperial examination. In the river of Chinese culture, which has been flowing slowly for thousands of years, the floods of globalization carrying modernity are pouring in like raging mountain torrents. The impact is so enormous that even the most stable rocks of the river bed cannot avoid the erosion or cracks. The dramatic changes of the image of contemporary Chinese countryside have made everyone realize that the reshaping of Chinese countryside by modernity has been inevitable reality. There is no way for people in the countryside to go back to their good ol' days.

The French lyric poet Baudelaire became an early literary critic who elaborated on the concept of modernity in aesthetics through his observations of Paris during the Second Empire. Baudelaire argued that "By 'modernity' I mean the ephemeral, the fugitive, the contingent, the half of art whose other half is the eternal and the immutable." *(Charles-Pierre Baudelaire. The*

象征与讽喻

—— 现代性中的挣扎

    在当代的城镇化大潮中，我们也需要意识到这样一个问题，那就是这个时代的变革和以往任何一次相比都是更为深刻和具有颠覆性的。西方社会在中世纪以后在接二连三的文艺复兴、宗教改革、启蒙运动的思想变革中不断地被重塑，即使一直处于变化的过程当中，当面临资本主义时代所催生的现代性时，西方社会仍然面临着巨大的困惑。反观中国社会，直到鸦片战争以前，中国文化倚赖于唐宋以来不断加强的儒学思想的主导地位，倚赖于宗家体制和科举制度的制度保障，一直保持着稳定的文化结构。在中国文化这条千百年来慢慢流淌的河床里，全球化的洪水裹挟着现代性有如山洪暴发般的泄入，再稳定的磐石也很难保证不留下一些冲蚀或裂隙。中国当代乡村面貌的剧变让所有人明白现代性对于中国乡村的重塑已经是一种现实，乡村的人们也再回不到曾经的岁月静好。

    法国抒情诗人波德莱尔（Baudelaire）通过对第二帝国时期的巴黎的观察，成为在美学上对现代性思想较早进行阐述的文艺批评家。波德莱尔认为"现代性就是短暂的、易逝的、偶发的，它是艺术的一半，而艺术的另一半则是永恒的和不变的。"[1]波德莱尔所生活的巴黎，正是法国工业革命达到高潮且城市资产阶级大量出现的时代，城市面貌和当代中国一样发生着现代化的巨变。在法国的主流社会眼中，旧日

Painter of Modern Life and Other Essays. 1967. P13) Baudelaire lived in Paris at a time when the French industrial revolution was at its peak and the urban bourgeoisie was proliferating. It was a time the urban landscape was undergoing the same dramatic changes of modernization as contemporary China. In the eyes of the French mainstream, the idylls of the old days were no longer appropriate for the modern city. The public wanted a thorough urban renewal to cut off the tie with the classical era. This was a modern reshaping of urban space characterized by boulevards. The new Paris of the industrial age was revealed with wide boulevards that ran through the city. Baron Haussmann almost completely wiped off the medieval urban texture of Paris in order to rebuild the city's road system. A large number of new "Haussmannism" houses formed the façade of the boulevards. The image and lifestyle of a medieval city that had developed in Paris over centuries were completely replaced by a new bourgeois urban landscape in a matter of twenty or thirty years. For Baudelaire, modernity was full of endless hostility to the past. In Paris of nineteenth century, the ephemeral and transient perception of modernity was hastily trying to bury the ancient myth of the eternity and immortality of the classical age. But everything came so fast that when modernity destroyed the classical age, the ruins of the classical age pierced out of the wrapping of modernity and tore at it in the same way. Being in that era, Baudelaire revealed the sense of order that came with modernity, and meanwhile felt the sense of fragmentation brought by it. So the poet rarely praised the city in his poems, instead he revealed another side of the city through characters as poets, bohemians, prostitutes, and chiffonniers, who wandered on the edge of the city and the masses. Benjamin, a German literary critic, discovered this image of the "Chiffonnier" in his analysis of Baudelaire's poems about the city of Paris. *(Walter Benjamin. The Arcades Project. Edited by Rolf Tiedemann. Translated by Howard Eiland, Kevin McLaughlin. 1999. P349)*

Here we have a man whose job it is to pick up the day's rubbish in the capital. He collects and catalogues everything that the great city has cast off, everything it has lost, and discarded, and broken. He goes through the archives of debauchery, and the jumbled array of refuse. He makes a selection, an intelligent choice; like a miser hoarding treasure, he collects the garbage that will become objects of utility or pleasure when refurbished by industrial magic.

In Benjamin's view, the "aura" that surrounded the works of the classical era, whether in art or in urban architecture, had disappeared in the mechanical reproduction of modern arts, and the mysterious and complete artistic experience had been lost. Modernity is like a powerful cleanser. Despite the glamorous look, the artworks spawned in modernity had their time dimension unfortunately cleansed by it, and lost their traditional "originality"(Echtheit in German) - that is, the characteristic of an artwork created in a particular time, environment, and context. In the city, under the neat and orderly appearance of modern urban space, the vitality of the city seems to have been sucked away. What still remains the vitality of the city is the congestion and noise of the crowd left over from the last century. Likewise, modernity has washed away the haze that once covered the Chinese countryside. The image of "distance" has dissipated. As the aura of traditional arts is fading, its corresponding symbolic representation also seems inopportune in this era.

的田园诗不再适合现代都市。大众希望通过对于城市的彻底改造来与古典时代进行切割。这是一次以林荫大道为特征的现代性城市空间的重塑。工业时代的新巴黎展现在人们眼前的是贯穿城市的一条条宽阔的林荫大道。奥斯曼（Haussmann）男爵为了重建巴黎的道路系统，几乎将巴黎的中世纪城市肌理完全抹去。大量新建的"奥斯曼风格"的住宅构成了林荫大道的街景，巴黎多个世纪以来所形成的中世纪城市面貌和生活方式在短短二三十年之内被新的资产阶级格调的都市风貌和生活所取代。在波德莱尔看来，现代性对于过去充满了无尽的敌意，在十九世纪的巴黎，现代性的短暂与瞬时的观感正匆忙地试图埋葬掉古典时代的永恒与不朽的古老神话。但是一切来得太快，现代性摧毁了古典时代，古典时代的废墟也从现代性的包裹中穿刺出来，同样地撕扯着现代性。身处那个时代，波德莱尔揭示了现代性来临的秩序感，却也同时感受到了现代性带来的破碎感。所以诗人在他的诗集中少有对城市的赞美，而是透过诸如诗人、波西米亚人、妓女、拾荒者这一类游走在城市和大众边缘的人物展示了城市的另一面。德国文学批评家本雅明（Benjamin）在分析波德莱尔关于巴黎城市的诗作时，发现了这个"拾荒者"的形象。

　　这里有这么个人，他的工作就是每日在首都拾荒。他将这个大城市所拒绝的、丢失的、抛弃的、损毁的一切东西收集并分类。他检视这些道德败坏的档案，和杂乱的垃圾堆。他做出选择，一个智慧的选择。就像一个守财奴看护着自己的财宝一样，他将这些垃圾聚敛起来，因为只要被工业魔术师加以翻新，这些垃圾可以成为有用之物或是令人喜悦的东西。[2]

　　在本雅明看来，不管是艺术作品，还是城市建筑，弥散在古典时代的作品周围的"光晕"在现代艺术的机械复制中消逝了，神秘而完满的艺术体验失落了。现代性像是一剂强力清洗剂。在现代性中催生的艺术作品，看似光鲜，却被现代性清洗掉了时间的维度，失去了传统艺术作品的"本真性"——也就是作品在特定时期、特定环境、特定语境中被创造出来的那种特性。在城市中，在现代性城市空间的整洁有序的表象下，城市的生命力却好像被抽走，仍在维持城市那丝活力的只是上个世纪残留下来的人群的拥塞和喧闹。同样，现代性也将曾经遮罩在中国乡村之上的那层烟霭冲刷干净，"远"的意象消散了。随着传统艺术的光晕的消逝，其所对应的象征的表征方式也在这个时代显得不合时宜了。

　　而波德莱尔透过躲在人群之外的拾荒者的眼睛，发现在城市的角落或是隐秘之

Baudelaire, through the eyes of the chiffonniers hiding out of the crowd, found there buried the dreams of history in the corners or some secret places of the city. The "Flaneur" paused in the bustling crowds of the big city and lingered there to contend against the fleeting nature of modern society. Through their gaze, the flaneurs discovered the fragments of modernity under the veil of allegory. They collected these fragments and reexamined them. From Baudelaire's poetry, Benjamin noticed a means of observation with a melancholic quality - "Allegory." "Baudelaire's genius, which is nourished on melancholy, is an allegorical genius. For the first time, with Baudelaire, Paris becomes the subject of lyric poetry. This poetry is no hymn to the homeland; rather, the gaze of the allegorist, as it falls on the city, is the gaze of the alienated man." *(Walter Benjamin. The Arcades Project. Edited by Rolf Tiedemann. Translated by Howard Eiland, Kevin McLaughlin. 1999. P10)* Allegory differs from one to one correspondence between signifier and signified advocated by Saussure, instead it reveals obvious differentiation between signifier and signified, or another implication beneath the signification. In his book "The Origin of German Tragic", Benjamin studied allegory as a literary form, arguing that allegory is more inclined to pursue the authenticity of the times than symbols, while this authenticity is often times dark and desolate, as opposed to heroic glory of classical symbols. Allegory is also fragmentary. Stripping away the false appearance of totality, fragmentation is the intuitive experience of the subject. For Benjamin, the world of "modernity" is a world of ruins. As he said, "Allegories are, in the ideological realm, what ruins are in the material realm".

Talking about the reshaping of the Chinese countryside by modernity, village-in-city is perhaps the best example of Chinese villages struggling in modernity. As more and more moved in, houses are seen taller and taller in these villages once located on the edge of the city. The original farm houses with earthen walls and tiled roofs have been replaced by four or five-story buildings, often with small spires covered by glazed tiles protruding from the roof covering the stairwell. The houses were built as close to the boundary of the villagers' homestead as possible. The exterior walls of neighboring houses are almost against each other. All the daily needs of urban life can be satisfied here. The first floor of the buildings is all kinds of shop fronts for migrant workers and poor students to linger: convenience supermarket, internet cafes, small restaurants, beauty salons, small inns, and newspaper stands. Above the second floor, there are partitioned flats for rent, where young and vigorous men and women make their humble abodes. Under the cover of urban life, the villagers, who are in the minority of population and have lost their farmland, yet manage to maintain their habits of what they consider as village life. The village women would prefer to move their washing machines out of the house to do laundry together where they can gossip. Life here is a like a mixture of fish and dragons, mud and sand, yet rich and colorful. The village-in-city is miscellaneous, therefore booming and polyphonic; it is changing violently, therefore full of desires. It is an independent kingdom. Modernity strives to remodel these villages, yet failed to integrate it with the urban blocks right across the street. Instead these villages have become a novelty that is different from both traditional villages and contemporary cities. It is like a kind of "historical garbage" despised by "planners", but it stubbornly hoards itself. Village-in-city has witnessed the occurrence of "modernity" in China, but it has not become a monument of "modernity". On the contrary,

处，恰恰可能埋藏着历史的梦想。"游荡者"在大城市熙熙攘攘的人群中停顿下来，驻留下来，抗衡着现代社会的转瞬即逝。游荡者通过他们的凝视，发现讽喻遮蔽下的现代性碎片，他们收集这些曾经的碎片，重新审视它们。从波德莱尔的诗歌中，本雅明注意到了这一种具有忧郁气质的观察手段——"讽喻"。"波德莱尔的天才，被忧郁滋养着，这是讽喻体的天才。在波德莱尔那里，巴黎第一次成为抒情诗的主体。他的诗不是家乡的赞美诗；相反，却是一位讽喻家的凝视，而当他的目光落向城市，这是一位局外人的凝视。"[3] 讽喻不同于索绪尔（Saussure）所认为的能指和所指之间的一一对应关系，而是在能指和所指之间揭示明显的差异、或是表意之下的另一层寓意。本雅明在其著作《德国悲剧的起源》中研究了讽喻这一文学形式，认为讽喻相较象征而言更为追求时代的真实，而这个真实往往是黑暗的、荒芜的，而非古典主义的象征那样笼罩着英雄的光辉；讽喻也是碎片性的，它剥去总体性的虚假表象，碎片化才是主体的直观体验。在本雅明看来，"现代性"的世界是个废墟化的世界，如他所说"讽喻在思想领域就如同物质领域的废墟"。

回头看现代性对于中国乡村的重塑，城中村或许是中国乡村挣扎于现代性中的最好案例。曾经在城市边缘的村子，随着越来越多的人口的移入，这里的房子逐渐在长高，原先的土墙瓦屋的农居已变成四五层高的建筑，突出屋顶的楼梯间部位还经常盖以琉璃瓦的小尖顶。这些村民自己建造的楼房尽量贴近着自家宅基地的界线，相邻住户家的外墙几乎已贴在一起。这里的城市生活设施一应俱全，楼房的一层都是让打工者和穷学生流连忘返的各种门面：小超市、小网吧、小饭馆、美容美发店、小旅店、书报摊。二层以上则是供出租的房屋，年轻的血气方刚的男男女女们蜗居在这里。在城市生活的遮掩之下，占少数的村民们虽然已经失去了耕地，但仍维持着他们以为村居生活的习惯，村妇们也宁可把家里洗衣机搬出来也要扎堆在一起洗衣，聊家长里短。这里的生活鱼龙混杂，泥沙俱下，却又丰富多彩。城中村是芜杂的，因而又是浑厚的和多声部的；它是剧烈变化着的，因而也充满着种种欲望。这里是一个独立的王国。现代性努力地改造着这些曾经的村子，但它总不能和马路对面的城市街区融为一体，而是成为一种既迥异于传统乡村又有别于当代城市的新奇之物。它像是一种"规划者"所不屑的"历史垃圾"，但是它固执地自我储存起来。城中村见证了"现代性"在中国的发生，但它并没有成为"现代性"的丰碑，相反"现代性"在此断裂，形成了碎片和废墟。

受到政府保护的传统村落和在废墟中野蛮生长的城中村是中国乡村的两个极端

"modernity" has broken here and split into fragments and ruins.

Traditional village preserved by the government and village-in-city growing wildly in ruins are two extreme images of the Chinese countryside. What is in between are millions of ordinary villages. Modernity for most of the ordinary villages is not just a concept of the present, but should be understood as a process of collapsing and ruins piling up. Any historical fragments that have survived are the expression of the compromise between modernity of many different periods in the past. For such villages, the primary task is perhaps the historical "disenchantment" of the rural image - the removal of the historical aura that envelops the rural image. In this process, the formal norms of history - the modernity of the past - should be respected and not become the shackles of the modernity of the present.

意象，在它们中间是几百万个普通的村子。现代性对于大部分的普通乡村而言不只是一个当下的概念，而是应该被理解为一个现代性不断坍塌而废墟不断堆积的过程，所幸存下来的任何历史片段都是过去诸多不同时期的现代性相互妥协之后的表现。对于这样的乡村，首要任务或是乡村意象的历史"怯魅"——祛除笼罩在乡村意象上的历史光晕。而在这个过程当中，历史的形式规范——即过去的现代性，既应受到尊重，也不应成为当下的现代性的桎梏。

[1] BAUDELAIRE C. The Painter of Modern Life and Other Essays[M]. London: Phaidon Press, 1967: 13.
[2] BENJAMIN W. The Arcades Project[M]. Edited by TIEDEMANN R, Translated by EILAND H, MCLAUGHLIN K. Boston: Harvard University Press, 1999: 349.
[3] 同 [2]: 10.

Phenomenon & Dictionary

—— Dictionary of Rural Phenomenology

"Withered vine, old tree, crows at dusk; Tiny bridge, flowing brook, and cottages; Ancient road, bleak wind, bony steed. The sun sinking west, a heart-torn traveler at the end of the world." In this short lyrics "To the Tune of Tian Jing Sha - Autumn Thought" by the Yuan Dynasty composer Ma Zhiyuan, the 28 Chinese characters are used to sketch ten scenes and one person, representing the bleak state of mind in autumn. The first three stanzas of the lyrics list nine seemingly isolated things in two-character word groups, a rhetorical technique known as "Liejin" in ancient Chinese poetry, which means to lay out objects of aesthetic significance directly to the audience, and usually these objects have no priority. In each line of this poem, a scene is presented by the former two words as environmental elements and the latter word as element of living creature. But looking at each stanza individually, it is difficult to say that the elements of living creature and the environment form a relationship of subordination. The three stanzas constitute three disjointed scenes, from the near to the distant "sunset" and "heart-torn traveler", thus forming a panorama. However, this panorama is not coherent like a long scroll of traditional painting, but rather like a montage commonly used in film footage. This seemingly simple rhetorical device completely ignores the grammatical structure of subject, predicate, and object. But each of the selected fragments seems to have the function of self-discourse, and when put together, these fragments can autonomously organize themselves into an imagery structure without the help of grammar.

This rhetorical device of "Liejin" is more or less a reflection of the Chinese attitude towards

现象与辞典

—— 乡村现象学辞典

　　"枯藤老树昏鸦，小桥流水人家，古道西风瘦马。夕阳西下，断肠人在天涯。"在这首元代散曲家马致远的小令《天净沙·秋思》中，短短二十八个字，却勾绘出十景一人，展现了秋天的萧瑟心境。小令前三句每两个字一组，罗列了九个看似孤立的事物，这是中国古代诗歌中一种被称为"列锦"的修辞手法。列锦简单来说就是把具有美学意义的事物直接铺陈给观众，通常这些事物也无主次之分。在这首小令每一句的景物铺陈的过程当中，都由前面两个词作为环境要素和后面一个词作为活物要素构成一个小景。但是单独看每个短句，也很难说活物要素和环境要素构成一种主从关系。三个短句构成三个不连贯的小景，然后从近推到远处的"夕阳"和"断肠人"，从而构成一幅全景图。但这幅全景图不是像传统书画长卷那样连贯的，而更像是电影镜头常用的蒙太奇画面效果。这种貌似再简单不过的修辞手法完全无视主谓宾的语法结构，但经过挑选的每一个碎片式的事物似乎都有自我言说的功能，而且这些碎片放到一起的时候还能完全不需要借助语法而自我组织起一个意象结构。

　　列锦的这种修辞手法或多或少地映射出中国人对于采用何种言说方式以形成思想文本的态度，也就是相比严谨的语法结构，并不排除碎片式的言说也是一种有效的方式，甚至认为这种需要被言说对象自我意会来最终完成的方式是更高级的一种。

what form of discourse to choose to constitute the ideological text, that is, it does not exclude fragmentary discourse as an effective way compared to those strict grammatical structure, or even considers the form that requires self-interpretation of the addressee to finalize the discourse a more advanced way. In contrast to the rigid structure of the Western classical era that stresses on order, the naturalistic tendency of Chinese inspired by Taoism has made the Chinese people less bound by rigid structure in all aspects other than politics, especially on the aesthetic level, leaving much room for flexibility in the structure. Therefore, Chinese culture is most inclusive. In this cultural context, the historical fragments of rural image torn apart by the wave of urbanization are like pebbles disintegrated from a huge rock in the river and taking shape through continuous scouring and grinding. It is impossible for the pebbles to be restored to the previous rough rock, yet they are waiting to be discovered and reassembled into some new buildings.

Ancient Heyang village in Jinyun county, Zhejiang province, provides an experimental sample of rural linguistics. Heyang is a historical and cultural protection area on provincial-level, which is well preserved from spatial pattern to architectural form, and is a typical example of the traditional villages in this area. But the value of rural studies of Heyang lies not only in Heyang itself, but also in the series of ordinary villages of various sizes along Heyang's big stream, which are undergoing the same changes as other common villages found anywhere in China and are the real state of the transforming Chinese countryside. The continuity of ancient village traditions and the penetration of the outside world resist and compromise each other. In the midst of this great change, the village of Heyang seems like an old man who has suddenly had aphasia because of the "shock". However, if we patiently talk with this old man, there might be no strict grammar in his words, but he can still reveal some contradictory fragments of vocabularies to us: a courtyard full of barren grass, a couvert at the entrance of an alley where elderly gather, an old moon pond surrounded by new village houses, a cluster of horse-head walls which is hidden by the new buildings and no longer able to outline the skyline. These spatial and symbolic fragments disintegrated under the erosion of the times have lost their formal gloss, but their connection with the deep structure of traditional rural aesthetics makes them settle down slowly and turn them into independent entries concealed under modernity. In the contemporary design context, these entries shall not become banal formalities, stereotypes or clichés, but rather a rural dictionary that needs to be rewritten.

Reduction of phenomenology may be a method to interpret these entries that are gradually losing or have already lost their synchronic meaning. In order for the observer to focus on the conscious analysis of the self, the method of Epoché is introduced into the reduction of phenomenology to force the object to reveal itself by putting all the predetermined existence of the object into brackets, neither negating nor affirming it. Shao Yong has a similar argument in "Chapter of the Inner Observation", "By observation, it is not that we shall observe with eyes. We observe with heart, as opposed to eyes. Then we observe with reason, as opposed to heart." That is, a process from suspension (Epoché) to eidetic reduction, and further to transcendental reduction. As a design proposition, the phenomenological reduction of rural

和西方古典时代的那种讲求秩序的严谨结构相比，由道家根源所引发的中国人对于自然主义的倾向，使得中国人在政治以外的各个方面，尤其在审美层面，并不太被严谨的结构所束缚，而是在结构中留下了很多的灵活空间，所以中国文化也是最能包容的。在这样的文化语境中，被城镇化浪潮所撕裂的乡村意象的历史碎片就像是一块巨大的岩石在河床上裂解并被不断冲刷打磨而形成的鹅卵石，这些鹅卵石并不必然更无可能去还原到先前的那块粗糙的岩石，但是它们也正等待着被人发现并被重新组装到新的建筑体当中。

浙江省缙云县的河阳古村提供了这样一个乡村语言学的实验样本。河阳古村是省级历史文化保护区，从空间格局到建筑形制都保护得较好，是这一带乡村传统风貌的典型。但是河阳的乡村研究价值不仅在于河阳古村自身，而在于沿着河阳大溪直到镇上的一连串大小不一的普通村落，而这些村子正经历着和中国任何一个地方的普通乡村一样的变化，是真实的转变中的中国乡村的状态。古村传统的延续和外部世界的渗透在这里互相抵抗和妥协，身处这场巨变之中，河阳的乡村好像一位因"震惊"而失语的老人。但是如果耐心地和这位老人对话，在他的只言片语中或许已经找不到严谨的语法，但是他仍能向我们揭示出一些矛盾的词汇片段：长满荒草的天井、老人围坐的巷口廊棚、被新农居围绕的老月塘、天空被新房子遮住而不再能勾勒天际线的马头墙群。这些在时代的侵蚀下而解体了的空间和符号的碎片失去了形式的光泽，但是和传统乡村审美的深层结构的勾连使得它们慢慢沉淀下来，成为轶失在现代性下的一个个独立的词条。在当代的设计语境中，这些词条不应该变成一种套语、陈规或是陈词滥调，这是一部需要重新书写的乡村辞典。

如何对这些逐渐或已经失去共时性意义的词条进行释义，现象学的还原或许是一个可以使用的方法。在现象学还原中，为了让观察者能将注意力集中到自我的意识分析之上，利用悬置的方法，将对象的所有预先设定的存在放入括号之中存而不论，既不否定也不肯定，从而逼使对象显现自身。邵雍在《观物内篇》有异曲同工的论述，"夫所以谓之观物者，非以目观之也。非观之以目，而观之以心。非观之以心，而观之以理也。"也即从悬置到本质还原，再进而到先验还原的一个过程。作为一个设计学的命题，这里比不打算将乡村词汇的现象学还原深入到先验的领域而以至于过于抽象而无一丝形式可循。这里将使用的现象学还原是一种类似于红酒品鉴的活动，认识主体的信念并没有被完全悬置，只是让其更为专注于最为核心的体验。贝克韦尔（Bakewell）在其著作《存在主义咖啡馆》中以咖啡为例解释了现象学的一般还

vocabulary is not intended to go deep into the transcendental field that makes it too abstract without a clue of form to follow. The phenomenological reduction to be adopted here is an activity similar to wine tasting, where the beliefs of the cognitive subject are not completely suspended or only to an extent by which the subject can better focus on the core experience. Bakewell explained the general problem of phenomenological reduction in her book "At the Existentialist Café", taking coffee as an example. *(Sarah Bakewell. At the Existentialist Café. 2016. PP40-41)*

What, then, is a cup of coffee? I might define it in terms of its chemistry and the botany of the coffee plant, and add a summary of how its beans are grown and exported, how they are ground, how hot water is pressed through the powder and then poured into a shaped receptacle to be presented to a member of the human species who orally ingests it. I could analyse the effect of caffeine on the body, or discuss the international coffee trade. I could fill an encyclopaedia with these facts, and I would still get no closer to saying what this particular cup of coffee in front of me is. On the other hand, if I went the other way and conjured up a set of purely personal, sentimental associations - as Marcel Proust does when he dunks his Madeleine in his tea and goes on to write seven volumes about it - that would not allow me to understand this cup of coffee as an immediately given phenomenon either...

Instead, this cup of coffee is a rich aroma, at once earthy and perfumed; it is the lazy movement of a curlicue of steam rising from its surface. As I lift it to my lips, it is a placidly shifting liquid and a weight in my hand inside its thick-rimmed cup. It is an approaching warmth, then an intense dark flavour on my tongue, starting with a slightly austere jolt and then relaxing into a comforting warmth, which spreads from the cup into my body, bringing the promise of lasting alertness and refreshment. The promise, the anticipated sensations, the smell, the colour and the flavour are all part of the coffee as phenomenon. They all emerge by being experienced.

If I treated all these as purely "subjective" elements to be stripped away in order to be "objective" about my coffee, I would find there was nothing left of my cup of coffee as a phenomenon - that is, as it appears in the experience of me, the coffee-drinker. This experiential cup of coffee is the one I can speak about with certainty, while everything else to do with the bean-growing and the chemistry is hearsay. It may all be interesting hearsay, but it's irrelevant to a phenomenologist.

The rhetorical device of "Liejin" provides a good reference for us when we make phenomenological reduction. "Liejin" does not involve any action rendered by subject or object. It only allows the subject's will to be projected onto things in order to make the phenomena of things appear. Take "To the Tune of Tian Jing Sha - Autumn Thought" as an example again, the poet projected his own state of mind onto the things he chose, so that things such as "vine, tree, road, horse" appeared as phenomena of "withered, old, ancient, bony". When attempting to make a phenomenological reduction of Heyang's rural entries, such as Dao Tan (the name of the courtyard in Heyang): We can certainly depict it with a lot of architectural descriptions

原问题。

那么，一杯咖啡是什么？我可以用它的化学成分和咖啡的植物学来定义它，并附加一份总结来说明咖啡豆是如何种植和出口，如何被研磨，热水是如何被压滤过咖啡粉末再被冲入容器中，然后被端给以口进食的人类中的一员。我可以分析咖啡因对人体的影响，或者讨论咖啡的国际贸易。我可以用这些事实填满一部百科全书，但我仍然说不清我面前这杯咖啡是什么……

相反，这杯咖啡是浓郁的香气，是可以即刻感受到的泥土气息和浓香；是从咖啡表面旋转升腾的蒸汽的懒散运动。当我把它举到唇边，它是平稳流动的液体和我手中的那个厚杯子里的一份重量感。这是一片逐渐靠近的暖意，然后是我舌头上的强烈而浓黑的味道，从一丝朴素的颤动开始，然后舒缓到一种惬意的温暖，从杯子扩散到身体，带来持续的提神和精神焕发的承诺。这个承诺，预期的感觉，这个气味、颜色和味道都是咖啡现象的一部分，它们都是通过经验而显现。

如果我将这些都作为纯粹的"主观"要素而剥离，以使我的咖啡显得"客观"，那我会发现我这杯咖啡 —— 也就是说，以我作为喝咖啡的人经验来看 —— 没有任何现象留存。这杯经验式的咖啡是我可以肯定地去谈论的，而别的任何和咖啡豆生长和化学有关的一切只是道听途说。这些可能都是有趣的传闻，但与现象学家无关。[1]

列锦的修辞方式为我们做现象学还原时提供了很好的参照。列锦不牵涉到任何的主体或客体的动作，只是让主体意志投射到事物之上以显现事物的现象。还是以《天净沙·秋思》为例，诗人将自己的心境投射到他所选的事物之上，从而让"藤、树、道、马"等事物呈现出"枯、老、古、瘦"的现象。对河阳的乡村词条尝试进行现象学还原，比如道坛（河阳一带对院落的称呼）：我们当然可以对它进行大量的如尺寸、比例这样的建筑学上的描述，但是悬置这些客观的描述，静静地站在低垂的檐下待上几分钟，然后突然感受到昏暗的屋檐所裁剪的一方天光或许就是道坛的本质。以最敏锐的直觉来体验这个空间的存在本质，这个现象也是心灵意向性的表达和延伸。而对于乡村的这种心灵感受，不同的人一定有不同的体验，但是集合起来看，也很难否认存在着某种集体记忆的移情作用，也就是说乡村审美的深层结构或多或少会主宰这个心灵的意向性。所以，河阳乡村词条经过现象学的还原之后得到的原型并不是完全抽象的逻辑原则，而是具有某种主观意识的，是具有某种原始潜能的，

such as dimensions and scale, but let's suspend these objective descriptions and stand quietly under the low-hanging eaves for a few minutes. Then suddenly we feel that it is the sky light cut by the dark tile eaves that may be the essence of Dao Tan. To experience the existential nature of this space with the sharpest intuition is also an expression and extension of the intention of the mind. Different people must have different experiences of this spiritual feeling of the countryside. However, it is hard to deny the existence of a certain empathic role of collective memory when examined collectively, which means that the deep structure of the rural aesthetic will more or less dominate the intention of the mind. Therefore, the archetype obtained from Heyang's rural vocabulary through the phenomenological reduction is not completely an abstract logical principle, instead it has some subjective consciousness, some original potential, and of course, aesthetic significance.

The interpretation of these rural entries is not simply a matter of establishing the relationship between the signifier and signified. The historical meaning cannot and does not have to be forever fixed in the signifier of the symbols. In order to avoid the interference of the obsolete discourses from the past architectural historical research and gain some freedom of thought in the process of writing, there are different ways of writing in the process of interpreting these rural entries. Literature texts describing the countryside in past years, static images that freeze time, and phenomenological sketches of subjective expressions, all these ways of writing refer to and intertwine with each other, and are liberated from its original social or psychological structure and put into a new intertextual context of free dialogue between various texts. A narrative of ordinary daily life, a scene of a corner of the house, a rural image of intuitive experience, these signifiers keep passing and weaving in this intertextual context. The past texts can interact with the future texts at any time to produce new meanings. In this new intertextual dictionary, no single discourse is dominant, and the relationship between signifier and signified is ambiguous. Each entry is not fixed as a symbol, but to be rewritten by the users.

当然也是具有美学意义的。

　　对这些乡村词条的释义并不是简单地建立能指与所指的关系，历史意义并不能也没有必要被永远固定在符号的能指之中。为避免过往建筑历史研究中陈旧话语的干涉，也为了获得自我书写时的一些思想上的自由，在这些乡村词条的释义过程中可以有不同的书写方式。过去年代描写乡村的文学性文本、静止的让时间停滞的影像、主观表达的现象学速写，这些书写方式互相引征和交织在一起，把各自从原先的社会或心理结构中解放出来，投入一种新的各类文本自由对话的互文语境中。一段普通日常的生活叙事、一个房前屋后的角落场景、一幅直觉体验的乡村图景，这些能指在这个互文性的语境中不停地传递和编织，过去的文本可以随时与未来的文本交互而产生新的意义。在这部新的互文性语境的辞典里，没有一种话语占据主导地位，能指和所指之间的关系是扑朔迷离的，每一个词条作为符号而言都不是固定的，而是有待于被使用者重新书写。

[1] BAKEWELL S. At the Existentialist Café[M]. New York: Other Press, 2016: 40-41.

第二章

河阳

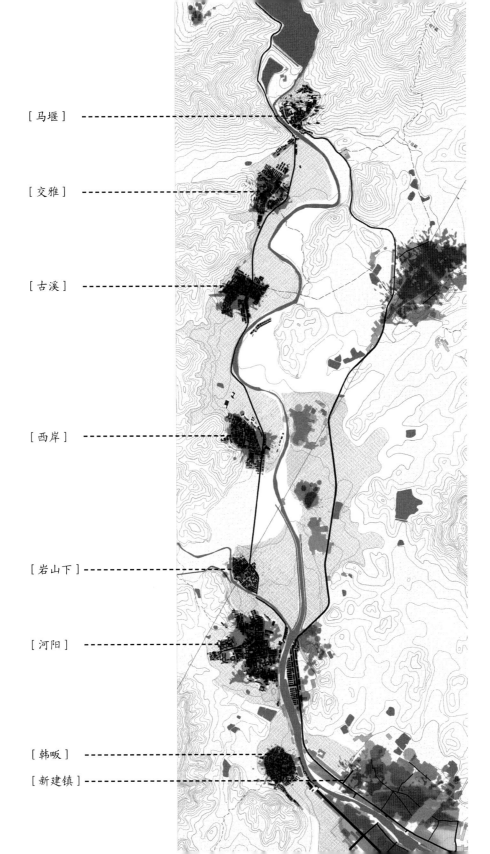

［马堰］

［交雅］

［古溪］

［西岸］

［岩山下］

［河阳］

［韩畈］

［新建镇］

[地理]

　　河阳古村，地处浙中南腹地缙云县。

　　缙云一地为括苍山一脉，颇有仙气。《隋书·地理志》有云："括苍有缙云山"。而地名又因山名而来，唐代《元和郡县图志》有云："因山为名"，"缙云山，一名仙都，一曰缙云，黄帝炼丹于此。"此处山色常见赤色，是典型的丹霞地貌。而许慎《说文解字》有注解："缙，帛赤色也。"古人以赤多白少为缙，以缙云名山最是恰当不过。缙云一地，山水形胜，其中尤以仙都鼎湖峰为名。山水诗鼻祖谢灵运赴任永嘉太守途中经过缙云，乘竹筏走好溪水路，见缙云仙都山水，写下了这样的感叹："漾百里之清潭，见千仞之孤石。历古今而常在，经盛衰而不易。"仙都西南方向几十里处则是群山之间难得的平原之地，虽不是沃野千里，但水源丰沛，颐养一方百姓。所以，缙云一带既有文人隐士倾心向往的险绝山水，又有布衣百姓赖以生存的鱼塘良田，为古时极好的避世之地。

[源流]

　　唐朝末年，原吴越国掌书记朱清源朱清渊兄弟俩为避五季之乱，而迁居此处，即缙云山往西北约二十五里地处，也即现在的河阳。因朱氏先祖源于河南义阳郡（现为信阳），故将此处取名"河阳"。

　　河阳朱氏聚族而居，人丁兴旺，素有"烟灶八百，人口三千"的称号。随着考取功名出仕或外出经商，河阳朱氏也有不断外迁者，近者有青田、永嘉、台州等地，远者到福建、台湾。自明洪武年间初修谱序，因枝繁叶茂，历代朱氏勤修家谱，到民国共有老谱十六部，是传统家族传承研究的宝贵资料。

[堪舆]

　　河阳之地山环水抱，风水极佳，是远离纷争的山水宝地。据《义阳朱氏家谱·祖居纪原》，"形家者曰：其地为五龙抢珠。古屋逆朝西南，维直作正厅三进，翼以重厢，大门外有荷田数亩，前坑后溪，横束如带，左右夹沙各二，合中陇。五龙又有外缠，右缠为大小岩山，左缠为二井山，内水口为黄碧山，又名下臂山，外水口为潴头岩。水收三十余里，重堤绵亘里许，以荫住居，内乐为东溪山，外乐为屏风山，龙沙缠堤，森林合抱。前有道院、尼庵，后有梵宫、精舍。山明水秀，代毓人文。"大约意思是朱氏祖屋，也即河阳村所在之处，是五龙抢珠的风水格局。该地倚靠仙霞岭余脉中峰山为主的五座连绵的大小山体，右边是大、小岩山，左边是二井山；同时后面紧靠河阳大溪，形如玉带横束；而且周围佛寺道观，钟灵毓秀。

[宗族]

　　中国传统社会的血缘关系以姓氏传承，并往往制定严格的宗法制度以巩固和凝聚这种关系，祠堂正是这种宗族力量的象征和核心，也是宗法制度的浓缩和物化。河阳也不例外，且族内支派众多，《义阳朱氏家谱·祠记》中就记录有朱大宗祠、圭二公祠、文翰公祠、虚竹公祠、荷公特祠、圭六公祠、丹崖公祠、哲六十七公祠、恒三公祠、玉天公祠、有周公祠等十五座祠堂。这些祠堂根据不同的祭祀对象而各有不同：有祭祀开源始祖的总祠，如朱大宗祠；有各房派的支祠，如圭二公祠、虚竹公祠；有祭祀乡贤的特祠，如荷公特祠（孝子祠）；还有后辈为感念先祖之间的人情义气而建的忠祥遗绩祠；以及十分罕见的女人祠堂 —— 信女祠。

　　河阳朱氏以宗族规范约束子弟品行，订立了一整套刚柔并济的家规二十条附小规十条，以忠孝仁义的儒家纲常对婚丧嫁娶及日常行为作了明确的规定。其中，尤以耕读为先，家规第七条"务耕读"明确提出："耕读，人生正业。子孙志诗书者，宜先品行；子孙服田亩者，宜勤力作。"河阳朱氏子弟耕读持家，出仕者众。

[宅巷]

　　宋元两朝朱氏一门八进士，元代在村里立了八士门以纪念之。村中主街即对着该门，以此主街为中轴线，形成了一街五巷的基本格局。之后因河阳人丁旺盛，主街两侧出现商铺、医馆等，颇有一些繁华的景象。

　　具有当地特征的村宅"十八间"大都在垂直于主街的五条巷子内。浙中一带民居以"十三间"为主要形制，河阳"十八间"多出五间，从一个侧面也说明当地之富庶。河阳民居的外墙常采用青砖，并有马头墙高低错落，门额、门罩、窗罩等处常采用砖雕。入口处常采用匾额，以凸显家风。内部木构、门窗则采用大量东阳木雕工艺。历经明清两朝，河阳古村也逐渐发展为当地规模最大、结构最为完整的村子。目前村中尚有百余栋旧宅，计 1500 余间，大多为明清两代所建，是宗族聚居的古民居建筑群。

[产业]

　　河阳古村四周农田与水塘汩罗交织，刚解放时有耕地四千余亩，随着建设用地逐年增加，耕地已减少到目前的一千余亩。当地农业一直以水稻为主，解放后蚕桑产业有很大发展。当地还盛产麻鸭，且历史悠久，离河阳不远的上游西岸村自明清就有专门售卖鸭卵的店铺。

河阳地处两港汇合之处，为通衢之地。二十世纪六十年代之前，每逢四、十，都有周围乡村及山民来赶市，其中有批发杉树、毛竹、大米的集行，也有"怡和堂"药铺，热闹非常。甚至在民国时期，当地合作社还印制过角票、分票使用于市。河阳子弟重农经商，一直生活富足。清末曾有号称"全国第二富"，仅略逊于同时期胡雪岩的虚竹公曾买下苏州城一条街。当地也一直流传着"有女嫁河阳，赛过当娘娘"的民谣。

[河阳村落群]

河阳边上大溪，又称新建溪，属山溪性河流，系钱塘江水系源头。据《义阳朱氏家谱•公济桥记》，"河阳碧山，傍有溪，上为保居堤，下为关浸堰，实西乡南北之通津，其源发于雪峰，北经夏家畈，汇黄连坑、历山、茭岭诸水，自七、八两都，并六都至河阳、复汇插花岩、杨桥、章岭下，南东之水，互注而来。虽非洪流巨浸，亦众水之会归处也。"

新建溪自河阳村西北流入，从村东流过，自西向东穿越村庄，往下到新建镇。新建也是一个千年古镇，历史悠久，以前为永康辖地。元朝二次毁于兵祸，明初重建后，改称新建。从新建镇沿新建溪溯溪而上，里许就有一个村子，递次会经过韩畈、河阳、岩山下、潘村、下杨、西岸、东岸、古溪、交雅、马堰等村，直到马堰边上的白马水库。

这里的新建盆地是缙云县最大的盆地，也有群山之中难得的大片良田，温暖湿润，水源充足，是典型的浙中乡村地区。这里既有水乡平原的生活特征，又因为靠近山区，也带有山村的简朴气质。河阳古村是这一带乡村风貌的典型，但是河阳的乡村价值不仅在于河阳古村自身，而在于沿河阳大溪这一连串大小不一的村子。这些村子没有河阳这样的规模，但形态格局各有特点，形成一个合而不同的古村群落。

虽然河阳古村早在 2000 年就被列为省历史文化保护区，2013 年"河阳村乡土建筑"又被列入第七批全国重点文物保护单位，但是在城镇化的大潮流中，曾经偏安一隅的河阳古村也很难再独善其身。三里之外新建镇的范围不断膨胀，乡村人口也不断向小镇集聚，离小镇最近的韩畈村已基本和新建镇连成一片，貌似迟早会被划入新建镇的建设范围之内。在当代的城镇化进程当中，保护一整片地区的乡村风貌的重要性绝对不亚于对单一古村的保护。当然这两者也是相辅相成的，河阳村落群的保护发展需要河阳这个典型的引领，而河阳古村的生命力也是依附于整个河阳村落群的生命力而存在的。

河阳村落群的研究，沿着主要的村道选取韩畈、河阳、岩山下、西岸、古溪、交雅、马堰七个村展开。

韩阪

[韩畈]

　　出新建镇，沿河阳大溪往西北方向的第一个村子即是韩畈。

　　韩畈建村已约九百余年，相传黄姓二十一公从江西逃居缙云，其子孙一部分迁居此地，故韩畈村多姓黄，村内有黄氏宗祠。此地的黄氏先祖商定，凡黄姓住地挖井两口，井边栽树，表示黄姓住地。又因古籍有载：双井有树即为韩也。此地有井两口并有树，村前有一片田畈，故名韩畈。

　　韩畈村主种水稻，也产茶叶，村民或去外地养麻鸭。或许因为不如河阳朱姓人多势大，韩畈村民一直喜欢耍拳弄棍，村里汉子几乎没有一个不会一点拳脚功夫的。据说当时与河阳村争田水打架，大大的河阳村打不过小小的韩畈村。

　　韩畈村现有500多户人家，1300多人，因最为靠近集镇，村内民居密度很高。随着新建镇的扩张，韩畈几乎已经和新建镇连成一片，大量传统民居已经被拆除并新建为五六层的水泥楼房。

河阳

[河阳]

　　河阳古村，历史悠长，源流有序。宗谱记载吴越国朱氏兄弟为避五季之乱，而徙居缙云。此地右靠大溪，左依大小岩山，山水形胜，自此朱氏宗族世居于此。因原籍河南信阳（也称义阳），当地朱氏一直以义阳朱氏自称，村名则取名"河阳"以彰显其本。

　　河阳村内以八士门街为中轴，在元代就形成一溪两坑的水系格局和一街五巷的村庄布局。村内民宅错落有致，普遍宽厅多间，以"十八间"闻名，居住条件舒适。村中尚有十余座明清古祠堂、百余栋旧宅，计1500余间，大多为明清两代所建，是宗族聚居的古民居建筑群。

　　义阳朱氏历代祖先以耕读传家，重农经商，人才辈出，富甲一方。当地一直流传"有女嫁河阳，赛过当娘娘"的民谣。

岩山下

[岩山下]

　　岩山下村，在河阳村西北里许，是因大小岩山下面的田畈而形成的小小的自然村，村民多为赵氏。岩山下村和河阳村一大一小，仅只一条小溪之隔。因为距离相近，村子又小，解放之后岩山下村和河阳分分合合，现在又属于河阳行政村。

　　岩山下村虽然只有80多户、400余人，但也有悠久的历史，且亦以单一姓氏为主，村内有赵氏祠堂。村边岩山上有岩山殿，奉文昌帝君。

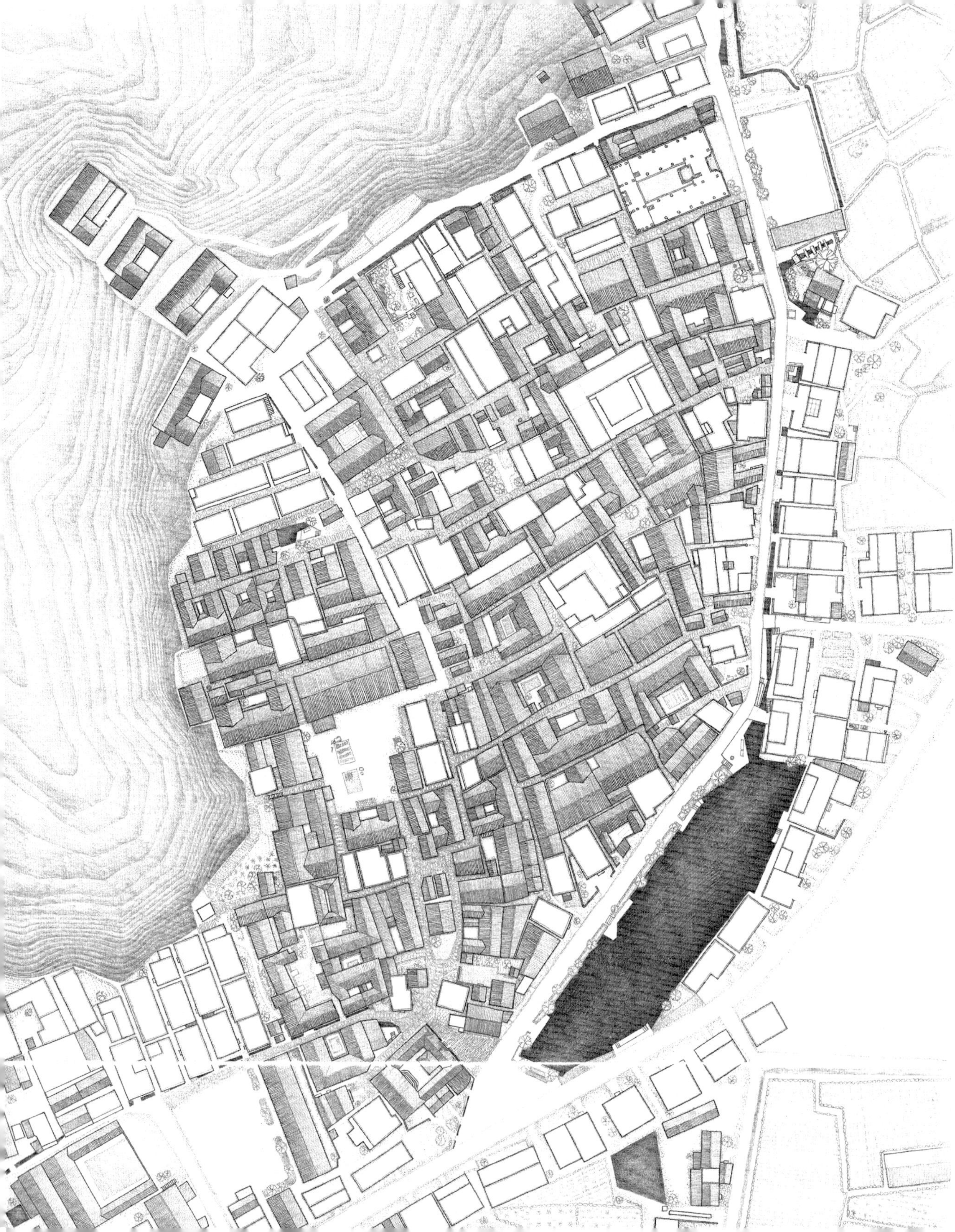

西岸

[西岸]

　　西岸，古称西皋，因村庄建于河阳大溪西边的高地之上而得名。相应的大溪对过还有东岸村，但规模稍小。村西北角和村道交接处有大水塘一围，名曰后堰潭，旁有清代道光年间所立碑记，平日潭周围常有村民闲坐。

　　西岸村内 300 余户，1000 余人，以吕姓为盛，也入迁最早，相传是南宋理学大师吕祖谦的嫡长支后裔。吕祖谦之玄孙吕伯良在元初迁居西岸后不久即开始大兴土木，建造自家宅院。西岸出去的吕氏一脉，甚至在闽台开枝散叶，颇为繁盛。村内现仍有吕氏宗祠。

　　村内农户大量养殖麻鸭，明清以来就开有专门出卖鸭蛋的商店，曰鸭卵店。由此推断，缙云麻鸭的历史起码在六百年以上。鸭卵店至今尚存三间明初古屋，在缙云为已经发现的最早古建筑之一。

[古溪]

　　古溪，村如其名，河阳大溪流经该村东侧，有名山靠于后，佳水绕于前，山水之势不亚于河阳古村。溪水通过明渠暗渠被引入村子，山村内处处方塘。常常是三四户村宅围绕水塘而建，形成一个个小小的邻里单元。

　　村中赵姓人为主，赵氏祠堂即在村中心位置。祠堂、空地、方塘合在一起形成古溪村的中心，村道贯穿村中心而过。

交椅

[交雅]

　　交雅村，在当地大寒尖岭山脚。相传葛洪寻访炼丹之地，途经交雅问道于黄姓老翁，老翁指近道于葛洪从交雅大寒尖岭脚上山直至大寒尖。
　　可惜的是该村在 2009 年曾发生过一次严重火灾，共烧毁房屋九十余间，所以村内除了少数几个合院，几乎都是近二十年内新建的农民房，整齐的行列式宅基地上是四五层高的小炮楼，这几乎就是当代浙江农村风貌的缩影。

马堰

[马堰]

　　马堰，以村子边上的白马水库而得名。河阳大溪，现名新建溪，就从这白马水库下泄。村子就在水库出水口左近，另一侧是山，地势在这里陡然升高，犹如堰坝上的一个村子。
　　马堰村甚少农田，人口本就不多，四五百人不到，现在更是过半村民在外谋生。村内为数不多的传统村宅或是空关或是失修，但是也有不少新建的小炮楼样式农居房沿村道逐级升起。

第
三
章

河
阳
乡
村
辞
典

居第目

[道坛]

　　推开宅门，站在门房里，面对的是一个双披屋檐围绕的四合院，乌黑浓重的瓦片屋檐从天空裁剪出一片天光引入天井。

　　这一带有轩辕氏祭祀的世代传承，每年有清明和重阳的春秋二祭，各地宗祠乃至自家天井内都会设坛祭拜。而温州、丽水一带，又把门前院内的庭院或天井称为"道坦"，不知河阳乡言是否受此影响，这里的天井被称为"道坛"，无形中就好像带了些道家仙气。受这里的道家文化影响，这里的村民喜欢在道坛四角砌花坛种植花卉或堆条石摆放盆景，尤其是剑兰，当地有"中堂天官赐福，道坛剑兰天竹"的说法。

　　现在的道坛不一定规置得整齐，有些甚至长满杂草显得有些荒凉，但依稀仍有旧时光的印记。石砌花坛虽已残破不堪，里面的剑兰仍在。天井一角连着放在一起的几口大水缸，里面也仍有几尾金鱼在游动。竹竿撑起的晾衣架上偶尔也会有几件衣服挂在那里。母鸡们在杂草中四处觅食，老人们则坐在小竹椅上打发时间。

　　大一点的人家在宅子两侧还有厢房和伙房，也围成小天井，和中间的道坛形成"目"字形的格局。穿过天井的穿廊在宅子两侧都有便门，走到小巷当中再稍微走几步，又是另一户人家的侧门。乡村之中往往路不拾遗，门都是开着随便进出的，小孩子们一溜烟就可以跑到人家的道坛里玩耍。

　　这些道坛或大或小，一个连一个地被编织到一起，形成一块巨大的内外空间连绵不断的居住肌理包覆在这块土地上。

[Courtyard of Tao]

　　Pushing the door open and standing in the gate room of the house, one faces a courtyard surrounded by double eaves. The dark tile eaves cut out a piece of sky to lead the light into the courtyard.

　　There is a tradition of sacrifice to Xuanyuan(Chinese common ancestor in Taoist mythology) in this area. On Qingming Festival in spring and Chongyang Festival in autumn, altars would be set up in ancestral halls or even in one's own residence. In Wenzhou and Lishui, inner courtyard or front patio is called ''Dao Tan''. Whether or not influenced by that dialect is uncertain, though the courtyard here is called ''Dao Tan'' as well, except the difference in Chinese characters meaning ''Tao altar'' exuding some Taoist fairy spirit. Affected by the Taoist culture, the villagers here like to build flowerbeds or place potted plants on piled slates at the four corners of the courtyard. Gladiolus is their favorite to secure the residence metaphorically, suggested by a local saying ''hang a portrait of heavenly blessing official in the central hall, guarded by the gladiolus and bamboo on the altar''.

　　Now the courtyards are not as neatly arranged as before, some of which even look desolate with weeds all over. The marks of time can still be traced. The flowerbeds are devastated, yet the gladiolus is still there. At a corner of the courtyard are the large terracotta pots that still collect the rainwater with goldfish inside. Clothes are sometimes hung on the bamboo racks. The elderly sit on small bamboo chairs to idle away the time, with hens strolling around in the weeds.

　　Larger houses may have wing rooms and kitchen adjoining both sides of the house from outside, which enclose small patios, forming a triple courtyard pattern with the courtyard of Tao in the middle. The corridor passing through the courtyard has access doors on both sides of the house. Walk into the alley and take a few steps further, there is the side door of another house. In the countryside, the doors are always open during the days and children can sneak into neighbor's courtyard anytime.

　　These courtyards, large or small, one by one, are woven together, forming a gigantic dwelling texture covering the country land.

原型的心灵意向：一方家庭生活起居的天光之井

Archetupal Intention: A space of family living filled with light

天井，干净、安静，春红晚白。

房子的中间留一块空地，人站在天井里，可以仰头观天，察天之气象，有繁星点点。或者，白云苍狗，风萧萧起于鱼鳞瓦片之上。

……

大院子不是天井，院子里一间一间生活起居的部落，才构成天井，是私人生活的一部分。如果有一只路过的大鸟从高处看，人就仿佛陷落在天井的中间。天晴的时候，住在里面的人，从这一间房到对面那间房，可以走一条斜线，从天井径直穿过。可是下雨天不行，必须弯弯绕绕，走马廊沿。

……

天井是一篇构思巧妙的文章，厅堂、厢房、厨房，像字、词、句，次第铺展分布。厢房有窗，可以观春秋，有棵桑椹树，昨夜风雨，地上落一层浅浅的紫色果儿。天井里栽两棵枇杷树，亭亭如盖，枇杷挂得一树金。墙角还有一丛芭蕉，一簇簇雨花在天井水塘里盛开，屋檐口下雨天，雨水溅在脆叶上，清新悦耳。一簇簇雨花在天井水塘里盛开，屋檐口的水就哗哗流，升腾起淡烟，从瓦上跌下，一缕如线。

……

板桥先生说，人生得意处，莫过有『茅屋一间，天井一方』，修竹数竿，小石一块，便尔成局，亦复可以烹茶，可以留客也。月中有清影，夜中有风声，闲心消受。

——王太生·《蚕老枇杷黄》

89

形式的变化潜能一：
向上裁剪天光的室外空间

Formal Transformation Potential I:
Upward space tailoring the light

形式的变化潜能二：
连绵的道坛肌理

Formal Transformation Potential II:
Texture of continuous Courtyards of Tao

[映月]

  月亮总是给人最为浪漫和诗意的遐想。苏州沧浪亭内有一处月洞门，曰"周规折矩"。洞门上方的砖额一面刻着"折矩"，另一面刻着"周规"。古人行走皆有法度，《礼记·玉藻》有云，"周还中规，折还中矩。进则揖之，退则扬之。然后玉鸣也。"

  河阳也出富商巨贾，明末清初朱氏出外经商者众。乾隆年间村内有缙云首富之次子朱柏轩，也称柏轩二翁，在村内所建宅子别具特色，融入造园意境。入宅前的门巷中共有四重门，中间两重为月洞门，上方分别悬挂柏轩之子所书的"循规"、"映月"的匾额。

  所谓"映月"，须有些景致映在这满月之上。假山、莲池、古木、湖石、飞檐，园中的任一处风光被收于这月洞框景之中，都好像被一层月的光晕包围而不可触摸。由外入内，入月洞门由此世界入彼世界，犹如入满月桂宫。但即便被这一轮圆月剪出的只是一片白墙，那后面的深宅内却好像有无穷的景致。故此，见到这个月洞门，即已想见这个地方的不凡。

  黄昏时，从"映月"踏入古宅的门巷内，这里好像是被两头的月洞门所尘封的世界，而童年时乡村记忆的梦境不断从这圆月形的门洞中被抛射出来。

[Reflecting the Moon]

  The moon always brings the most romantic and poetic fantasies. In the Pavilion of the Surging Waves in Suzhou, there is a moon gate named "Zhou Gui Zhe Ju". The brick plaque above the moon gate was engraved with "Zhe Ju" (make right-angle turn at corner) on one side and "Zhou Gui" (taking circular route when turning back) on the other. Ancient Chinese strictly follow rules in daily behavior, as promoted in "Liji" (a Confucian sacred book about social and political philosophy), "a gentleman shall take circular route when turning back, and make right-angle turn at the corner, a gentleman shall lean as if making a bow when stepping forward and lift the body up when stepping backward, so that the jade ornament he carries would produce rhythmic sound along with the body movement."

  Heyang is known for wealthy businessmen. From the late Ming to early Qing Dynasty, a lot of members of the Zhu house went out to do business. During the Qianlong period, Zhu Boxuan, the second son of the richest man in the village, built a mansion in the village with unique style that integrated artistic conception of gardening. There are four gates in the alley before entering the house. The two in the middle are "moon gates", above which are the plaques engraved with "Xun Gui" (follow the rules) and "Ying Yue" (reflecting the moon) written by Boxuan's son.

  When called "Ying Yue", there must be some scenery reflected on this "full moon" gate. The rockeries, lotus pond, ancient trees, lake rocks, flying cornices, any scenery in the garden is reflected and framed by the moon gate, as if they were surrounded by the moon halo and untouchable. From exterior to interior, the moon gate is the entry from one world to the other, like entering the Osmanthus Palace on the moon. Even the framed scenery is just a white wall, there seems to be endless views behind. Seeing a moon gate is enough to imagine how extraordinary a place is.

  At dusk, when one enters into the alley of this ancient mansion through the "Ying Yue" gate, the space between the two moon gates is like a sealed world, from which the fragments of childhood dreams about the rural memories are constantly reflected back.

原型的心灵意向：象征意义的圆月以对彼世界的框景完成对此世界的呈现和引诱

Archetypal Intention: The symbolic full moon completes the presentation of the other world and the seduction into it by framing it

我们前行。又圆又大的月亮落到小巷口，堵住小巷，好像巨大的古代圆洞门。

她说：我们走到月亮里好吗？她的脸越来越皎洁。

我说：那是你的家吗？

她靠近我走。我觉得我们走到光源里。

月亮落尽，我指指乌黑的夜空说，这不够浪漫。

她说，这很好。

—— 吉明 · 《走失的女子》

照片上的背景部分，经过漫长岁月，留下泛黄的时间痕迹，还有一些幻影似的模糊斑点。衬托着人像背后的斑驳墙垣上，一个很大的圆圈，轮廓分明。

墙上大圆圈是一扇圆形的大门？我问自己。

正是我童年时代熟悉的月洞门，童话般的门。

是一扇门。

我出生的老屋里，厅堂前有一个石面铺砌的院子，足供我嬉戏奔跑。花砖墙跟前，并列着几只大水缸，用以承接檐下的雨落水。每一只水缸都比我高得多。我躲在水缸后面，与小伙伴捉迷藏。院子两侧，东西相对，各有一个月洞门。圆圆的像天上月宫吗？我常常想起故乡的老屋，为什么我总是记得那两扇大圆门呢？是因为它不同于普通的长方形门框，圆圆的像天上月宫吗？抑或是，两扇大门之间，有一块小小的天地，曾经是我骑竹马驰骋的所在？我常常想起故乡的老屋，它充满了我童年的回忆。

—— 何为 · 《照片上的童年》

形式的变化潜能一：月洞框景的变幻

Formal Transformation Potential I: Scenic variations in the moon frame

形式的变化潜能二：月洞的构成——重复、放大、错位、切半
Formal Transformation Potential II: The arbitrary composition of the moon frame - repetition, enlargement, dislocation, cutting

[炊烟]

　　河阳人家的灶间一般就在两侧的洞头房，这里靠道坛比较远，天光进不来，白天也是昏暗的。随着天色渐暗，里面愈加黝黑。但是往往到了最不见天光的时刻，哄地一丝火苗从炉灶里跃出，随着噼里啪啦的声音，柴火越烧越旺，农妇开始在灶间忙碌起晚饭。炉膛里的火光，灶头锅盖上的蒸汽，小孩们时不时地溜进来围绕着母亲讨要几口刚烧好的美食。这个时候，这个不起眼的黑黝黝的地方将乡村生活最精华的记忆凝固在这里。

　　灶台的烟道就靠着外墙，排烟有的时候就是在外墙开个洞，复杂一些的就会让烟囱伸出屋顶。"暖暖远人村，依依墟里烟"，炊烟自古以来就是村野田居生活的象征符号。炊烟是人间烟火，有人就得冒烟，不冒烟就是没人了。

　　虽然今时不同往日，但遥想河阳当年"烟灶八百，人口三千"时的黄昏：高高低低的烟囱耸出屋面，像一群放哨的卫兵守护着这些老宅子。黑色的连绵的瓦屋面在烟囱里喷薄出来的白烟和热气中摇曳和隐约，整个村子在炊烟中云遮雾绕，形成一幅天上人间的幻象。

[Curling Smoke from Chimneys]

The family stove in Heyang is often located at either side of the house where is the farthest point away from the courtyard with the skylight unable to penetrate and is dim even in the daytime. It grows darker as the sky darkens. But when it becomes the darkest, a glimmer of fire jumps out of the stove. With crackling sound of firewood, the fire is burning more and more furiously. The countrywoman begins to prepare dinner. Fire in the stove, steam hovering above the stove lid, and children occasionally slipping in to ask for a few bites of freshly cooked food from their mother, this is the moment at which the best memories of rural life is sealed by this dark and unspectacular place.

The flue of the stove is often built against the exterior wall, while sometimes there is simply a hole on the wall to ventilate. A more sophisticated approach is to build a chimney on the roof. "The distant village dimly looms somewhere, with smoke from chimneys curling in the air." (Tao Yuanming's poem "Back to country dwelling") Smoke from chimneys has been the symbol of rural life from ancient times. It is the smoke of earthly world. Where there is people, there is smoke, vice versa.

Although time has changed, we recall the dusk of Heyang hundreds of years ago with "eight hundred stoves, three thousand residents": Tall or low chimneys project out of the roofs, like sentries guarding these old houses. The black and extending tile roofs appeared to be swaying and looming in the white smoke and heat spewed out of the chimneys. The whole village was veiled by the smoke, creating an illusion of earthly heaven.

原型的心灵意向：一缕烟，一个家，高高低低的炊烟守望着村子的活力

Archetypal Intention: Under a plume of smoke, is a family, smokes curling up and down, guard the village's vitality

炊烟是房屋升起的云朵，是劈柴化成的幽魂。它们经过了火光的历练，又钻过了一段漆黑的烟道后，一旦从烟囱中脱颖而出，就带着股超凡脱俗的气质，宁静、纯洁、轻盈、飘渺。无云的天气中，它们就是空中的云朵，而有云的日子，它们就是云的长裙下飘逸着的流苏。

如果你晚霞满天的时候来到山顶，俯瞰山下的小镇，可以看到一动一静两个情景，它们恰到好处地组合成了一幅画面：静的是一幢连着一幢的房屋，动的则是袅袅上升的炊烟。房屋是冷色调的，炊烟则是暖色调的。这一冷一暖，将小镇宁静平和的生活气氛完美地烘托出来了。

女人们喜欢在晚饭后串门，她们去谁家串门前，要习惯地看一眼这家烟囱冒出的炊烟，如果它格外地浓郁，说明人家的晚饭正忙在高潮，饭菜还没有上桌呢，就要晚一些过去；而如果那炊烟细若游丝，若有若无，说明饭已经吃完了，你这时过去，人家才有空儿聊天。

炊烟无形中充当了密探的角色。

——迟子建·《暮色中的炊烟》

形式的变化潜能一：
屋顶上的卫兵

Formal Transformation Potential I:
Soldiers above the roof

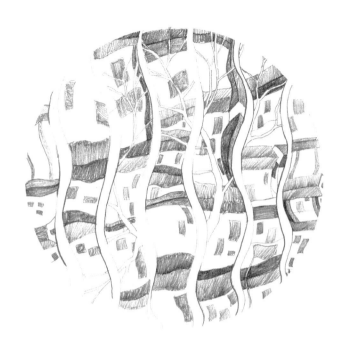

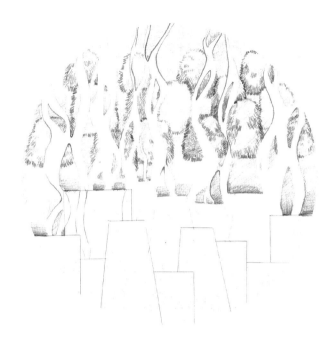

形式的变化潜能二：
炊烟后的村庄幻象

Formal Transformation Potential II:
Illusion of village behind the curling smoke

[户牖]

"凿户牖以为室，当其无，有室之用。"有了户牖，才有建筑空间的"有"与"无"、"实"与"虚"之间的转换。有了户牖，才有空气在空间的流通，才有光对于形体的塑造。

中国人讲门第，讲光耀门楣，入户门对于任何一户人家而言都是非常具有象征意味的。而且中国的传统民居只要条件允许一般都会采用内向性的封闭合院形式，入户门是建立居住仪式感的第一个环节，所以入户门尤为重要。以前大门一般时间是不开的，进出主要走的是侧门。侧门虽然要窄很多，但是稍有点家赀的，为了显示家族兴旺，侧门也会增加门罩，门罩就是用水磨青砖贴着墙面砌出来的装饰性挑檐，这个挑檐由斗栱、额枋、匾额、垂花等元素组合而成，像一顶大官帽压在门户之上，大大地增加了门户的尺度感和气势。门罩的这种做法和西方巴洛克建筑的大门其实颇有点异曲同工的意思。

但中国乡下的传统是耕读持家，如果只是一味显富不是太上得了台面的，尤其在河阳这个曾经出过八个进士的地方。所以入户门的匾额上往往是一些很有文人气的题名以显主人的清雅，如"竹崦"、"松台"；如"风月古"、"水云间"。

进到院宅内部，农村房子的窗牖比较朴素，最常见的就是柳条纹样的支摘窗。当然在一些大宅之中，也能见到复杂一些的纹样，或是还会结合一些民间故事的木雕。冬日的河阳，温柔的光线穿过窗棂照到屋内而在墙上或地上形成的阴影是最吸引人的。

[Openings]

"Cut out openings of doors and windows to build a room. Being empty, the room can be used as dwelling." (the Book of Tao and The by Laozi) With the doors and windows, it is possible to realize the spatial transformation between "being" and "not being", "actual" and "virtual". With the doors and windows, there is the ventilation in the space, as well as the light's shaping of the form.

Chinese attach great importance to the portal which refers to pedigree and bringing glory to the family in Chinese context. So the portal bears strong symbolic meaning to every family. Traditional Chinese residence generally adopts a spatial form of introverted enclosed courtyard if only the situation is allowed. The portal is where the first session of establishing a ritual sequence of dwelling, thus is of great significance. In the past, the main gate was closed at most of the time, while the side gate was used as the daily access to the house. Though the side doorway was much narrower, a wealthy house would have it covered by decorative cornice made with grey bricks against the wall to show their prosperity. This cornice is constituted with bucket arches, architrave, plaque, and floral-pendant decorations, which is shaped like a huge official's hat above the portal and significantly exaggerates its scale and stately manner. A comparison could be drawn between this practice and that of Baroque style portal in the west.

Farming and reading are the essential virtues of Chinese ancient scholars in the countryside, as opposed to the accumulation of wealth, particularly in place like Heyang which used to have eight Jinshi (successful candidate in ancient highest imperial examinations). Therefore, the inscription on the plaque above the portal here often used scholarly genre, such as "Bamboo Hill" and "Pine Terrace", or "Wind and Moon Encounters" and "Between Water and Cloud", in order to show the master's elegance.

Inside the courtyard, the windows of the rural house are relatively simple. The most common ones are willow striped awning windows. More complicated patterns, or wooden carvings about folk tales, can be found in some large mansions. In winter, with the gentle sunlight penetrating through the windows into the house, the shadow formed on the wall or on the ground is the most attractive.

卷一 屋宇 门楼

门上起楼，象城堞有楼以壮观也。无楼亦呼之。

卷一 装折 户槅

古之户槅，多于方眼而菱花者，亦遵雅致，故不脱柳条式。或有将栏杆竖为户槅，斯一不密，亦无可玩，如根空仅阔寸许为佳，犹阔类栏杆风窗者去之，故式于后。予将斯增减数式，内有花纹各异，后人减为柳条槅，俗呼『不了窗』也。兹式从雅，者亦可用也。

卷一 装折 风窗

风窗，槅棂之外护，宜疏广减文，或横半，或两截推关，兹式如栏杆，减在馆为『书窗』，在闺为『绣窗』。

卷三 门窗

门窗磨空，制式时裁，不惟屋宇翻新，斯谓林园遵雅。工精虽专瓦作，调度犹在得人，触景生奇，含情多致，轻纱环碧，弱柳窥青。伟石迎人，别有一壶天地；修篁弄影，疑来隔水笙簧。佳境宜收，俗尘安到。切记雕镂门空，应当磨琢窗垣；处处邻虚，方方侧景。

——计成·《园冶》

107

108

形式的变化潜能一：　　　┌
入户墙上的窗牖

Formal Transformation Potential I
Entry articulated by windows

形式的变化潜能二：
窗户纹样在墙面的扩展

Formal Transformation Potential II:
Extension of window patterns on the wall　　　┘

[墙垣]

　　徽派民居在江南一直有很大的影响，所以即使在浙中一带也还是能看到很多粉墙黛瓦的村镇。传统的房子是木结构的，但是作为围合结构的外墙一般以青砖居多。青砖由黏土烧结而成，以前大一点的乡村几乎都有自己的砖窑来烧青砖青瓦，这是旧时的工业化产品。青砖墙外面一般再挂白灰，时间久了，有些白灰剥落或透出后面的砖墙肌理，就很有历史感了。

　　在偏僻一些的乡村建房子，更多的就是靠山吃山，靠水吃水。在一些沿海或是山区有溪流的村子，常见的是石砌的房子。而在一些没有太多建材资源的地方，村民也时常用黄泥做版筑墙。当这些乡土材料所构成的单纯的墙垣肌理不断在一个村子里重复着，就形成了传统乡村最具有视觉冲击力的图像。

　　乡村宅院的墙垣形式是多样的，乡村工匠的审美能力也是不可低估的。走在河阳，随处可见不同的卷棚样式，相同的是那些曲线起戗优美得令人惊叹。这每一片墙垣不止是围合了宅院，它更是属于整个村子的，当走入某个空间稍微放大一些的拐角就会明白这层意思。在这里一片片不同样式的墙垣好像从房子独立出来，它们前后错落，左右生扶，在植物的掩映中自主地完成了一处新的具有诗意的室外空间构成。

[Walls]

　　Traditional residence of Huizhou style has been major influence in the area south of the Yangtze River. Even in the hinterland of Zhejiang province, villages and towns are characterized with white-wall and black-tile houses. Traditional houses adopt timber structure, while the exterior walls are mainly built by grey bricks. Grey bricks are sintered from clay. In the past, almost all large villages had their own brick kilns to burn grey bricks and tiles, which were the industrial products of the old days. White plaster was then applied to cover the grey brick wall, and flaked off over time revealing the brick texture behind. That was when the sense of history began to emerge.

　　Building houses in remote villages was more about relying on the environment. In coastal or mountainous villages with streams, stone houses were common. In places which lacked resources of building materials, villagers often used rammed earth to build the walls. When some simple wall texture formed by these vernacular materials repeats itself continuously, the most visually striking image of the traditional village is generated.

　　The walls of village houses have diverse styles. The aesthetic taste of the rural craftsmen can not be underestimated. Walking in Heyang, different curling gable styles are observed, with astonishingly beautiful curves. Every piece of wall not only functions as enclosure of one's house, but rather belongs to the whole village. It becomes evident when walking into an enlarged corner. Differently styled walls shaded by the plants are set back and forth, left and right, as if they were independent from the house, and autonomously form a poetic exterior space.

原型的心灵意向：外墙也是外部空间的内墙

Archetypal Intention: Exterior wall is also the interior wall of exterior space

卷三 墙垣

凡园之围墙，多于版筑，或于石砌，或编篱棘。夫编篱斯胜花屏，似多野致，深得山林趣味。如内、花端、水次、夹径、环山之垣，或宜石宜砖，宜漏宜磨，各有所制。从雅遵时，令人欣赏，园林之佳境也。历来墙垣，凭匠作雕琢花鸟仙兽，以为巧制，不第林园之不佳，而宅堂前之何可也。雀巢可憎，积草如萝，祛之不尽，扣之则废，无可奈何者。市俗村愚之所为也，高明而慎之。世人兴造，因基之偏侧，任而造之。何不以墙取头阔头狭就屋之端正，斯匠主之莫知也。

——计成·《园冶》

113

形式的变化潜能一：
一面山墙的随时修补

Formal Transformation Potential I:
Patching of a gable over time

形式的变化潜能二：
不同山墙的空间层叠

Formal Transformation Potential II:
Spatial layering of different gables

形式的变化潜能三：
一种山墙的无尽重复

Formal Transformation Potential III:
Endless repetition of gables of one kind

间

里

目

[浣渠]

　　河阳一带的村子外面是新建溪，上游就是白马水库，这里的溪水长年不断。溪水也流到每个村子变成一条条宽窄不一的小水渠，再流到很多人家门口。在有洗衣粉之前，这些小水渠是清澈的，渠里的水草随着水流而摆动。小渠上有很多的条石，踏着小水渠上的青石板进家门，很有跨过护城河的仪式感。这些条石横横斜斜、宽宽窄窄地在那里，形成一种特殊的韵律感。

　　家里终究抵不过门口的热闹，毕竟门口的洗衣石上才是村里各种新闻的集散中心，当然更是村妇们在操持一大家子的家务时可以找找乐子的场所。村妇们喜欢在同一个时间出来蹲在这里浣衣洗菜，这个时候就是村妇们"曲水流觞"的时刻。她们洗衣时因为家长里短而聊天打趣所达成的兴致，实在也是不亚于文人骚客们。

[Rinsing Ditch]

　　Outside of Heyang and other adjacent villages lies the Xinjian Stream, the upstream of which is the Whitehorse Reservoir. The stream runs all year round. It spreads into every village and becomes narrow or wide ditches, flowing to the doorsteps of the villagers' houses. When there was no washing powder, the water was clear with water-weeds seen swaying along with the current. Many houses in the village have stone slabs over the ditch as the doorsteps. Treading on the slabs entering into the house creates a feeling of crossing a moat. These slabs with varied width are set there in free angles, forming a special rhythm.

　　It is never as happening at home as out there at the doorsteps. These laundry slabs at the doorsteps is where the tabloids of the village are spread, and where women can have some fun after a day's busy work. Women would come out and gather here at the same time to do laundry or washing vegetables. This is a moment of village women's "floating wine cups along the winding water" (Wang Xizhi's famous social activity), at which their spirit aroused by gossip of tabloids while laundering is no less than that of the ancient poets.

原型的心灵意向：农妇家门口的"曲水流觞"

Archetypal Intention: Countrywomen's "floating wine cups along the winding water" kind of daily routine on their own doorsteps

120

当母亲又回到灶屋的时候，就会拉开后门，一溜小跑来到溪边，朝着黑暗里叫道：『别蹲在这里了，快点回家吧！』母亲的声音里总充满着无可奈何的埋怨情绪。于是，我拍拍清凉的水面，才慢慢地从洗衣石上站起来，往家里走去。逢到又要离开家了，我总要跑到小溪边，和那排青青的洗衣石默默告别，我的眼睛也会变得湿润起来，我甚至会像青蛙似地趴在湿漉漉的洗衣石上，在心底轻轻地呼唤着：洗衣石，你是我的梦，而我是你的梦吗？

……

从我记事的时候起，溪边的洗衣石便给了我欢乐。平平正正的洗衣石上，我和我的伙伴拿着石子玩一四七、二五八、三六九，谁输谁当新娘子。每当这个时候，蜻蜓也常常来凑热闹，在我们的耳边擦来擦去，一群群身条细细的长尾巴鱼在水里上下不停地穿行着时不时凑过来，啄着我们浸在水里的小腿，痒丝丝的，舒服极了。

当我稍大一点了，我便加入了婶婶嫂嫂的队伍，每天蹲在洗衣石上，洗菜、搓衣服。在这里，我跟着婶婶嫂嫂们学会了唱『的笃戏』，『何必大回娘家』我会唱，『孟姜女十二个月花名哭长城』我也会唱。阿千嫂是最会戏文的了。每当她一摆开唱戏的架势，我就会赶忙从水底捞起一块小石子，敲着洗衣石，为她当伴奏。

『阿千家里的，来一段「王千金法场祭夫」吧！』大六月里，溪边坐满了乘凉的人。不论是头发花白的还是穿着开裆裤的，只要见阿千嫂洗完东西，刚从洗衣石上站起，便会兴致勃勃地大叫。阿千嫂也不好推托，就会咧开嘴巴，露出一副好看的牙齿『林郎呀林郎呀』地唱起来。

——袁丽娟·《洗衣石，叫我怎能忘记你》

形式的变化潜能一：
曲水流觞

Formal Transformation Potential I:
Floating wine cups along the winding water

形式的变化潜能二：
洗衣石的空间变奏

Formal Transformation Potential II:
Spatial mutation of laundry slates

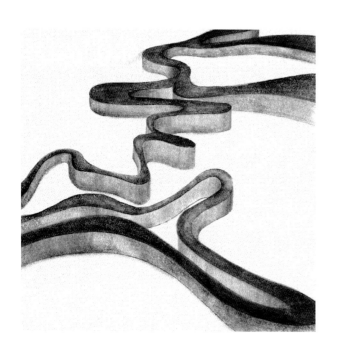

[村巷]

　　村子里的巷子和城市里的弄堂不太一样，没有那么的热闹。来的时候是个阴雨的天气，巷子里的卵石铺地有些湿滑。现在的人烟比以往少了很多，没有了人的踩踏，有些巷子大半已经被青苔和杂草占领。当然更可惜的是很多地方的路面随着乡村改造已经被水泥路面覆盖了。

　　这里的巷子的尽头往往有门洞，将巷子从村道切分出来，形成具有一定私密性的空间。巷子和道坛一起构成了最为基本的河阳一带的居住空间的图底关系。河阳的巷子与其说是家门口的过道，不如说是内部家庭空间的外延更为合适。十八间的大屋本来就有比较多的门，正面和两侧洞头房边都有门开向巷子。幽静的巷子里时不时地会有几只鸡从一边宅子的侧门扑腾出来，又从另一边宅子的侧门慢慢踱进去。可以想见当年顽童们也是一样在宅子间窜进窜出，而这之间的巷子必定也是他们玩耍的富于生气的好地方。

　　漫步在河阳的巷子里，每一步的感受是不一样的。巷子两侧随着围墙的变化为这个高高窄窄空间提供了丰富的空间切片。河阳是当地有名的生活富足的村子，巷子两侧都是十八间的大屋。外墙是已经斑驳的青砖粉墙，墙面上是大大小小的门洞和窗洞，以及它们上面高高低低的砖雕门罩和窗罩，合在一起为墙身增加了体感，变化着墙上的光影和肌理。周围有些村子没有那么富有，很多房屋要么是用石砌、要么是用夯土作为围护的墙体，巷子里不同材料墙体的交替却也让乡村的意味更加浓郁。也会有一些村宅的形制没有那么讲究，透过矮矮的围墙就能看到后面的道坛。巷子深处偶尔还会有被村民废弃了的房子的残垣，一些红花绿草从墙上的各处探出到巷子里来，很有一些"觉来村巷夕阳斜，几家，短墙红杏花"的意思。

[Alley]

　　Alleys in the village are not quite the same as those in the city. They are much quieter. On a rainy day, the pebbles on the ground are slippery. With much less residents than before, some alleys have been occupied by moss and weeds. It is more regretful that the pebble pavement in many places has been covered by cement with the rural reconstruction.

　　There is always a gate at the opening of an alley here, separating the alley from the village road and forming a private space. The alleys and the courtyards constitute a basic figure-ground pattern of dwelling space in Heyang area. An alley in Heyang is rather an extension of the interior family space than an aisle outside the house. There are relatively more doors in so-called Eighteen-room house, with opening to the alley in the front and both sides. Rosters would suddenly fly out of some side door and then stroll peacefully into the side door of neighboring house. It can be imagined that naughty kids used to dash from one door to the next as well. The alley in between must have been a lively place for them to play.

　　Strolling in the alleys of Heyang, each step brings different feeling. The variation of the walls on both sides of the alley generates varied spatial slices for this high and narrow space. Heyang is a village well known for its wealth and prosperity. On both sides of alley are the large Eighteen-room houses. The grey brick exterior wall is mottled. The doors and windows of various sizes, together with the engraved brick cornices above, enhance the stately sense of the wall, as well as the richness of light and shadow, and texture. The villages nearby are not as wealthy. Their houses tend to use more stone or rammed earth walls as enclosure, the alternation of which highlights the rural atmosphere. There are also houses that do not follow the residential form strictly, so that the courtyard is seen from outside. Deep in the alley, there are occasionally remains of abandoned houses. With flowers and grass growing out of the wall of the courtyard into the alley, it is quite a scene as the ancient poem puts "Roused up, see the sunset. Some houses, with red flowers on the low walls." (Xin Qiji's poem "Follow the genre of Huajian")

原型的心灵意向：将乡村生活的丰富切片串联起来的窄巷子

Archetypal Intention: A narrow alley connects seiries of spatial slices of varied rural life

高巷尽管风雨莫及，但又并非红尘不渡。仰头可以看到如沙的星，一线苍天瘦月，却已是有点像失群的云了。巷墙是村人屋壁的土砖，村人说它比础石还要坚硬，础石青苔滋生，容易剥蚀，而土砖干得发白，相传可以管上几百年。

百年以后如何，难道还用得着让我们去担心？说不定有能耐的子孙早已飘洋过海，哪里还会稀罕一椽木屋，记得这条小巷。足见，农民意识甚浓的村人，有时又似乎看得较远，想得较开了。巷里，来来往往，是世代在此种田的农人，或者为了让其便于在此安扎下去而在巷中走来走去的贩子。站在巷中，可以毫无遮拦地从敞开的门户窥见堂屋。牝鸡立几，未洗的碗箸狼藉地散满一桌，但堂屋黑色的地上仍扫得干净清爽，燕子排泄的粪便清楚可见。你问主人哪里去了么？巷中修桶的『圆木』会向你回答。是的，白天见不到劳力，主妇又挑水去了、上田栽菜去了，去塘边洗衣去了。但这些人，毕竟要常常来往巷中，所以巷仍然并不寂寥。你听过从巷石上传出的木屐声音么？那已经将巷的沉寂配上了一种有板有眼的古调。

——杨征宇·《村巷》

127

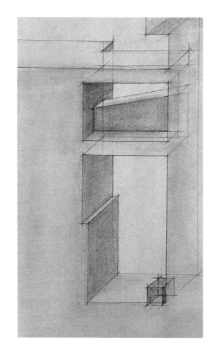

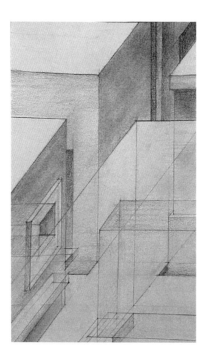
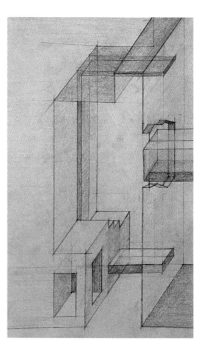

形式的变化潜能一：
窄巷内的空间切片

Formal Transformation Potential I:
Spatial slices in the narrow alley

形式的变化潜能二：
空间切片的组合

Formal Transformation Potential II:
Composite of spatial slices

[廊棚]

　　如果门口的青石板是村妇们的领地，那巷口的长条凳就是老人家的专属。河阳的巷口通常会夹一个凉棚，或者是一个过街楼，下面摆两条长凳子供村民闲坐。有遮风避雨的顶棚，这里自然就成为一个公共集聚的口袋空间，也是街巷内外空间的转换。村民们喜欢聚集到这里，尤其是闲在家里的老人们，家长里短，下棋打牌，这里就是他们的公共起居室。

　　我们走过外面的村道，看着老人们闲坐在这个巷口夹着的廊棚里，守着里面的门巷，而老人们也看着外面村道上的我们和人来人往。虽然简陋，但这处场所和园林中的亭子其实并无二致，观与被观在这里交织，这里是一个真正的看乡村市井风光的场所。

[Couvert]

　　If the blue stone slabs at the doorstep are the territory of the country women, the long benches at the entrance of the alley then belong to the elderly. The alley entrance in Heyang is usually covered by a couvert, or a gate-house, under which benches are placed for villagers to relax. With a roof shelter, it naturally becomes a pocket space for public gathering, as well as spatial transformation between the interior and exterior of the alley. The villagers like to gather here, especially the elderly who are idle at home. They chat and gossip, play chess and cards. It is their public living room.

　　We walked on the village road outside, watching the elderly sitting casually under the couvert at the entrance of the alley as if guarding the inside space behind, while they looked back at us on the road and other people moving around. Simple as the shelter is, there is no difference between this place and a pavilion in a garden. Spectating and being spectated are intertwined here. This is a place to see the real rural life.

原型的心灵意向：巷口带有遮蔽的口袋空间也是街巷内外空间的转换

Archetupal Intention: A sheltered pocket space at the entry of alley that transits between the interior and exterior

中午，船家送出酒饭来。傍晚到达塘栖，我就上岸去吃酒了。塘栖是一个镇，其特色是家家门前建着凉棚，不怕天雨。有一句话，叫作『塘栖镇上落雨，淋勿着』。『淋』与『轮』发音相似，所以凡事轮不着，就说『塘栖镇上落雨』。

———丰子恺·《塘栖》

他收拾收拾，再回到原先的弄堂口。那弄堂口多少有些阴暗，可是比较安定一些，过街楼避风挡雨，有一面墙根，可以堆放他的那些胶皮啊、鞋跟啊、钉子线绳，还有等着做的活计，或者做好等人来取的活计，也一并靠墙根。弄堂里的人，要么不来，要来就是一大堆，大大小小，男男女女，单的棉的，但都不是急等，所以就放在他这里，过一两天再来取。也不要领取凭证，不见得能认识人，可鞋总归认识的。而且，鞋这样东西，也不怕别人错领的。

———王安忆·《骄傲的皮匠》

133

形式的变化潜能一：
巷口的空间聚焦

Formal Transformation Potential I:
Spatial focus at the entry of alley

形式的变化潜能二：
口袋空间的移植

Formal Transformation Potential II:
Transplatation of pocket space

[簷基]

　　农村里一年到头都有农作物需要晾晒，在这里晒场被称为簷基，也就是铺簷子晒东西的地方。平常晒东西就是家里的道坛内或门口的空地上一摊簷子，或者是拿根三角杈的木棍作为一端，另一端支在窗台上支起两根横木棍，上面就可以放簸箕一类的晒具。如果上家里房顶方便的话，也有很多人家索性就把圆簸箕放在瓦屋顶上晒。到晒粮最集中的时候，簷基就非常紧缺了，村里人时常还会为簷基吵架甚至大打出手，这时候车来车往的车道一样不放过。

　　簷基的权属在村子里不甚明了，大约是靠近谁家或是谁先整出来的地谁就先晒，只是有个约定俗成的先来后到。有些村子就会开出一些大的公共簷基让大家晒粮。到了早上晾出和晚上收拢的时候，公共簷基就像是农村里的工厂，是最热闹的地方。偶尔遇上刮风下雨，就更加鸡飞狗跳一些。临时下雨要马上来抢收，如果刮个怪风把相邻两家的同类作物吹到一起而分不清彼此，还可能会引起主人间的争吵。深秋初冬的时候，老人们喜欢坐在这里晒太阳，同时照看着防止鸡鸭闯到簷基上，有的时候还要赶走在簷子上摔跤打滚的顽童。随着四季所晒的作物的变化，公共簷基上红的黄的绿的，好像一块随着四季变化颜色的大画板。

　　公共簷基是村子里少有的整块空地，这里最高光的时刻，可能就是当年放电影的时候了。银幕是竹竿撑起来的，下面是乌泱泱的人头，老老少少的心绪随着老式电影机投射到银幕上的光束的一明一灭和摇杆转动的哒哒声而起伏。

[Drying Yard]

　　In the countryside, grains need to be sun-dried all year round. In this area, the drying yard is called "Dianji", which means a place for laying out winnowing baskets to dry things. In most of the time, winnowing baskets are laid on the ground of the courtyard or any vacant land in front of the house, or sometimes on a simple device set up by two wood sticks supported by a trident wood stick at one end and the windowsill at the other. If it is convenient to go up on the roof, villagers would simply put the winnowing baskets there. When it comes to the busiest time of grain drying, it would be extremely difficult to find some drying yard. Quarreling or even fighting is not a rare scene at that time. The driveway would be taken as drying yard too, no matter whether there are vehicles running on it.

　　The ownership of a drying yard is not very clear in the village. It would not be wrong that whoever closer to the drying yard or is the first to clear it up should have the right of use. It is rather a social practice of "first come, first served". Large public drying yard would be cleared up in some villages. When spreading in the morning or collecting at dusk, the public drying yard is like a factory in the countryside and becomes the most happening place. If it rains suddenly, the situation would become total chaos when villagers rush to collect. If wind blows up the grains of different families and mixes them up, that would much likely lead to a quarrel. In late autumn and early winter, the elderly like to sit here in the sun, stopping chickens and ducks from jumping up to the drying yard, or sometimes driving the naughty kids away from wrestling and rolling on it. The color of red, yellow, and green alternates and mixes on the drying yard with the change of different grains, making it a big palette in drying season.

　　Public drying yard is a rare vacant ground in the village. The most highlighted moment is probably when an outdoor movie was played in the good old days. The screen was set up by bamboo poles. The mood of audience, whether old or young, fluctuated as the beam of light being bright and dim among the rhythmic sound of hand-rocker of the old projector.

原型的心灵意向：色彩与影像四季变幻的公共空地

Archetypal Intention: Public drying yard with rotated seasonal color and images

临河的土场上，太阳渐渐地收了它通黄的光线了。场边靠河的乌桕树叶，干巴巴的才喘过气来，几个花脚蚊子在下面哼着飞舞。面河的农家的烟突里，逐渐减少了炊烟，女人孩子们都在自己门口的土场上泼些水，放下小桌子和矮凳；人知道，这已经是晚饭时候了。

老男人坐在矮凳上，摇着大芭蕉扇闲谈，孩子飞也似地跑，或者蹲在乌桕树下赌玩石子。女人端出乌黑的蒸干菜和松花黄的米饭，热蓬蓬冒烟。河里驶过文人的酒船，文豪见了，大发诗性，说，『无思无虑，这真是田家乐呵！』

太阳收尽了他最末的光线了，水面暗暗地回复过凉气来；土场上一片碗筷声响，人人的脊梁上又都吐出汗粒。

——鲁迅·《风波》

每天清晨，太阳还没有出来，晒粉场上就忙碌起来。年老的妇女根据天边的云彩来猜度这一天的风向，然后调整一道道支架。支架的走向必须与风向交成十字，不然湿粉丝被风顺着一吹就会粘成疙瘩。马车辘辘地驶进晒粉场，接着一筐筐湿粉丝抬下来。洁白的，像雪一样纯净的粉丝悬在一行行架子上了，姑娘们赶紧伸手去摆弄它们。整整一天她们都要不停地忙活，用纤巧的手指去拆开纠扯到一起的粉丝，直到它们完全晒干，轻得像柳丝一样在风中徐徐飘动。

……

晒粉场上声音最高的就是闹闹了，她高兴干什么就干什么，有时还无缘无故地骂人。被骂的人从来不恼，都知道闹闹就是这样的脾性。她从电影上学会了迪斯科，有时就在沙子上跳开了。这时所有人的手里的活儿，喊着：『再来一个呀……』闹闹从来不听别人的话，她不想跳了，就一仰身子倒在热呼呼的沙子上，露出了雪白的肌肤。有一次她在沙子上躺下扭动着，说：『成天的，少个楞小子搂搂你了！』大家笑了。一个上年纪的妇女说：『就少个楞小子搂搂，那个楞小子恐怕还没生出来呢！』闹闹从沙土上跃起来，说：『真畅快呀，大家笑着，回过身子去摆弄粉丝了。姑娘们愉快地鼓掌……真畅快呀，

——张炜·《古船》

形式的变化潜能一：
簟子在空地上的固化

Formal Transformation Potential I:
Fixation of winnowing baskets in the drying yard

形式的变化潜能二：
簸子在空地上的减法

Formal Transformation Potential II:
Subtraction of winnowing baskets from the drying yard

[八士街]

　　每旬逢四、十两日为河阳一带的赶集日。记得第二次去河阳的时候，恰逢初四，又是周六，各种临时的摊贩将从新建镇直到韩畈村的公路两侧都占满了，四里八乡的都到这儿来赶集。平时稀疏，突然之间连绵两三里的路上人头攒动，车子是开不过去的。对于山里远途赶来采办生活必需的人而言，没人想错过赶集的日子。不过即使错过，有河阳在，终究是能弥补遗憾的。

　　河阳朱氏子弟以耕读传家，人才辈出，相传朱元璋听闻该地在宋元两代曾出八位进士，在村子正大门赐建八士门。门口一对无头的石狮子曰"稀罕"，取石狮无头而"稀罕"之意。八士门所对的村中主巷就称为八士街。河阳村民在外经商有方，富甲一方，乡村的市集在这里固定下来，形成一条寻常的村子很少见的商街，并且还将这条最初可能只是用来赶集的村道硬是发展成了能和古镇老街一较高下的规模。短短的两百米不到的街上店肆林立，南货店、药店、食品店、肉店、豆腐铺、饮食店、布店、裁缝铺、理发店、歇脚店，甚至金店，一应俱全。周围乡亲要采办一些时髦玩意，那是非到八士街来不可的。所以当地一直流传着"有女嫁河阳，赛过当娘娘"的民谣。

　　朱氏一族以出耕读出仕闻名乡里，过去每到阳春三月，还会在八士街上大开长廊宴，又称状元宴。八士街现在的境况当然大不如前，几乎难见到木排门全卸下迎客的店铺。沿街的店铺立面随着木排门的开开阖阖展现出里面不同的商业和生活场景，却也非常吸引人。

[Market Street of Eitht Scholars]

　　The fourth and tenth of every ten days are the market days in Heyang. For the second time we visited here, it was the fourth of the month and Saturday. Temporary vendors filled up both sides of the road from Xinjian town to Hanfan village, while villagers nearby were all out looking for their necessities. This country road which was quite empty on normal days suddenly became congested by one mile long stream of villagers that no vehicle could pass through. For those living in remote mountains, this is not some day that they would like to miss. But if they did, it is fortunate that Heyang is there to make up for their loss.

　　Farming and reading were held as the core virtues and passed down for generation in House Zhu of Heyang. It is said that when Zhu Yuanzhang, the first emperor of Ming Dynasty, heard that House Zhu had eight Jinshi (successful candidate in ancient highest imperial examinations) honored in Song and Yuan Dynasty, he bestowed the house to build the Gate of Eight Scholars at the entry to the village. There is a pair of headless stone lions guarding the gate called "rare", for a stone lion is rare without head. The main street facing the gate is then named Street of Eight Scholars. The villagers of Heyang used to go out to do business and became wealthy. A rural bazaar had been solidified into this commercial street that was rarely seen in other villages, and further developed to a scale that only the main street in some ancient town could match. Within a distance of two hundred meter long, different shops can be found including grocery store, drugstore, prepared food shop, butcher's shop, tofu shop, restaurant, cloth shop, tailor's shop, barber's shop, country inn, and even a gold shop. Those living in the neighboring villages had to come to the Street of Eight Scholars if they wanted anything in fashion. So there has been a saying in this area, "marry a daughter to a Heyang house, better than being a noble lady."

　　House Zhu was well known for making high officials in ancient time. In the past, a long table outdoor banquet, which is also called "Best Scholar Banquet", would be arranged on the Street of Eight Scholars on March to honor that achievement. The street is certainly no longer as busy as it was. It is hard to find a shop that is fully open with all front door panels removed. Behind the half removed door panels, commercial and life scenes are revealed and make a special street facade of countryside.

原型的心灵意向：乡村集市被固化下来之后的原初商业街

Archetupal Intention: Market street transformed from rural market

这街并不长，数起来不过四五十步。两边开着的店铺一共有十几家……有南货店，酱油店，布店，烟纸杂货店，药店，理发店，铜器店，鞋店，饼店……中间还夹杂着几家住家。

街的东头第三家是宝隆豆腐店，坐南朝北，两间门面，特别深宽，还留着过去开张时堂皇的痕迹。

——王鲁彦·《愤怒的乡村》

从运河边上的石码头上来，沿一条两边长满刺槐树的水泥路向前走，拐两个弯就是花街。一条窄窄的巷子，青石板铺成的道路歪歪扭扭地伸进幽深的前方。

远处拦头又是一条宽阔惨白的水泥路，那已经不是花街了。花街从几十年前就是这么长的一段。临街面对面挤满了灰旧的小院，门楼高高低低，下面是大大小小的店铺。生意对着石板街做，柜台后面是床铺和厨房。每天一排排拆合的店铺板门打开时，炊烟的香味就从煤球炉里飘摇而出。到老井里拎水的居民起得都很早，一道明亮的水迹在青石路上画出歪歪扭扭的线，最后消失在花街一户户人家的门前。如果沿街走动，就会在炊烟的香味之外辨出井水的甜味和马桶温热的气味，还有清早平和的暧昧。

——徐则成·《花街》

形式的变化潜能一：商铺门板的开阖节奏
Formal Transformation Potential I: Facade rhythm generated by removal of wood panels of shops

形式的变化潜能二：乡村集市的拱廊计划
Formal Transformation Potential II: Arcade plan of rural market

[古树]

　　从新建镇沿新建溪逆流而上探访河阳一带村落，没多远就是韩畈村。远远地就能看到韩畈村口一排排高大的樟树，下面围着两口方塘。韩畈村人多姓黄，祖上因为政治倾轧而迁居此地，恐有连坐而分散居住，规定黄姓住地挖井两口，井边栽树。几百年下来，这些古樟早已个个如大将军一般伫立在村口，古樟成林，冠盖如云。

　　沿公路往上十余里，交雅村口也是一片古樟林，古樟树下是入村的一弯圆拱石桥，桥下是村子的出水口。这里的古樟林更为密集，气势宏伟。

　　在乡村地区，村子的规模和地位也大约可从村头或村中的古树看出些一二。前人种树，后人乘凉，那一棵枝繁叶茂的大树必定意味着这里曾出过有见识的先人。

　　从村口这些古树下走出去求学或工作的人们回到村子，几里之外看到村口的古树，心中即是归家了。这几十年下来，乡村的风土景致大异，也或许只有这些古树还是当年的那个模样，还能唤起曾经在这里的少年记忆。

[Old Tree]

　　Going up along Xinjian stream from Xinjian town to visit the villages in Heyang area, Hanfan village comes into sight first. A row of tall camphor trees at the entrance of Hanfan village can be seen from a distance. There are two square ponds under the trees. Most families in Hanfan village are surnamed Huang, whose ancestors moved here because of political strife. They chose to live scattered being afraid of being implicated, and stipulated a rule that required Huang's residence should have two wells with trees beside in order to mark their surname. For hundreds of years, these ancient camphor trees have been standing at the entrance of the village, each like a great general. They have become a forest, and their canopies are joined together like clouds.

　　Going ten miles up along the road, there is another forest of ancient camphor trees at the entrance of Jiaoya village. Under the canopy is a stone arch bridge leading to the village, and under the bridge is the creek outlet. This camphor forest is even more dense and magnificent.

　　In the countryside, the size and status of a village could be judged from the old trees at the entrance of the village or inside. One generation plants the trees in whose shade another generation rests. An old leafy tree with large canopy means there must have been foresighted ancestors.

　　People left the village to study or work. When they return, it is when they see these trees by which they used to pass that they know they are home. The landscape of the village has changed greatly over the decades, perhaps only these old trees are still the same and able to evoke the memories of the youth.

原型的心灵意向：自然物所形成的承载了记忆和具有象征意义的场所

Archetypal Intention: Symbolic place bearing memories made by natural things

他两只手笨拙地拦过两颗熟透的西红柿，便飞一般地冲出了屋子。他没有去喂猪——让它暂且饿一会吧！他现在顾不得去喂它们了。他出了院门，下了公路，中学堂过小河，一口气爬上了村对面的山头。他大汗淋漓地坐在山顶一棵老杜梨树下，把上衣脱下丢在一边，一手拿着一颗西红柿，偏过来正过去地看着；用鼻子闻闻，在脸蛋上亲昵地擦擦。接着，不知为什么突然又蹦跳起来，光膀子举着两颗西红柿，绕着杜梨树热情奔放在跳将起来（很难说是舞蹈），直到一根裸露的树根绊了他一跤，才停止了这种疯狂行动。他嘿嘿笑着从地上爬起来，自己也为自己的行为害羞了，脸通红，赶忙朝四下里看看有没有人。

没人！正是中饭时光，山上劳动的人都回家吃饭去了。

他很不好意思地摇摇头，重新坐在老杜梨树下，眯起眼，出神地望着三伏天绿色浓重的高原，望着蓝天上的浮动的白云。啊，世界多好！他揩掉沾在西红柿上的土，想起了苏莹刚才对他说的话。

——路遥·《夏》

娶亲的人马在通过村子的时候，行进得特别缓慢——似乎为了让这热闹非凡的一刻，更深刻地留在村民的记忆里……巧珍骑在马上，尽量使自己很虚弱的身体不要倒下来。；她红丝绸下面的一张脸，痛苦地抽搐着。

在估计快要出村的时候，她忍不住用手捻开盖头一角……她看见了加林家的硷畔；她曾多少次朝那里张望过啊！她也看见了河对面一棵杜梨树——就在那树下，在那一片绿色的谷林里，他们曾躺在一起，抱过，亲过……别了，过去的一切！她放下红丝绸，重新蒙住了脸，泪水再一次从她干枯的眼睛里涌出来了……

——路遥·《人生》

形式的变化潜能一：
古树在村子各处

Formal Transformation Potential I:
Old tree dominating exterior space of a village

形式的变化潜能二：
古树在村宅内部

Formal Transformation Potential II:
Old tree dominating interior space of a house

祀

典

目

[祠堂]

　　农村是个血缘社会，没有什么事情是祠堂里不可以解决的，不管是操办婚丧嫁娶，还是执行家法族规。祠堂之于村民的重要性如同亚里士多德在《政治学》中所描述的城邦之于希腊公民。

　　河阳一带讲究耕读持家，以礼治人，所以祠堂兴盛，几乎每个村子都有好几处祠堂。清代缙云学训导余伟在河阳的《朱氏大宗祠记》中曾说："盖宗祠之由来久矣！于是以春露秋霜之感，将其爱存焘著之心，告慈，告孝，礼莫大焉！且因祀事，以会族属，群昭群穆，不失其伦。洞洞乎，属属乎，祖考如在其上，而子孙敬承于下。凡所为亲亲之道，胥于是乎在。"

　　河阳是个大村，朱氏一族分支众多，后世之中有显达者就开宗立祠。所以村子里有总的朱大宗祠，还有支派的圭二公祠、圭六公祠、文翰公祠、虚竹公祠等等，也有祭祀对象不为祖宗而是乡贤节妇一类的特祠如荷公特祠、信女祠等。

　　祠堂一般的形制大都为一进的合院式布局，上厅面阔少则三间多则五间，左右跑马廊，中间为天井。上厅的明间设有神龛，神龛上是历代祖宗的牌位，神龛后面的墙上是祖宗的画像。在比较大的祠堂，上厅对过一般是戏台，看戏是祭典仪式之后不可缺少的一项娱乐活动。从前的祠堂也是读书的地方，最为光耀门楣的莫过于勤读诗书考取功名，所以祠堂里平日里传来的都是琅琅书声。现在去到这些乡村祠堂，经常能看到老人们在厢房甚至就在上厅围坐在一起打麻将。作为一个乡村的公共场所，现在是无可厚非，不过以前估计是不敢的。

　　在从前官不能及的乡村这个小型复杂社会，祠堂围绕着宗族的凝聚力，它包容了信仰、裁决、娱乐、教育等各种乡村生活功能，是一个复杂的功能综合体。

[Ancestral Hall]

　　The village is a consanguineous society. There is nothing that cannot be solved in the ancestral hall, whether it is arranging weddings and funerals, or enforcing family laws. The importance of the ancestral hall to the villagers is similar as that of the city-state to the Greek citizens described in Aristotle's "Politics".

　　Ruled by Confucian ethics, and with farming and reading considered the essential virtues, almost every village has several ancestry halls. Yu Wei, the principal of the Jinyun school in Qing Dynasty, wrote in "Notes of the Main Ancestral Hall of House of Zhu", "There was a long history of building ancestral halls. People convey their feelings toward the change of seasons, show their respect to rites and their obedience to their parents in the hall. The house's inheritance, connection, and order are all maintained in this space."

　　Heyang is a big village with lots of branches of House of Zhu. Any descendant who had achieved great success would be entitled to build a new ancestral hall. So there is a main hall of House of Zhu, as well as lots of branch halls, such as Gui II Hall, Gui VI Hall, Wenhan Hall, Xuzhu Hall, etc. Except for ancestors, there are halls dedicated to virtues, such as Hegong Filial Piety Hall and Obedient Lady Hall.

　　The general layout of an ancestral hall is arranged around a courtyard, with an open hall of three bays or five, and galleries on both sides. There is a shrine in the hall, which is dedicated to the memorial tablets of ancestors of all generations. The portraits of the first ancestor are usually hung on the wall behind the shrine. In some larger ancestral halls, there is the stage across the courtyard. Watching drama was an indispensable entertainment after the memorial ceremony for the ancestors. The hall was also a place for schooling. The best way to bring glory to the family was to get an official rank through diligent reading and passing the national examination. The sound of reading used to be heard everyday in the hall. Nowadays, the scene has been changed to elderly gathering playing mahjong. It is now a common entertainment in the countryside, but no one dared to do so in the past.

　　In this small and complex rural society out of reach of the officials' governance, based upon the cohesion of clans, the ancestral hall contained faith, judgment, entertainment, education, etc. It was an actual complex with integrated functions in the countryside.

原型的心灵意向：增强宗族凝聚力的乡村功能综合体

Archetupal Intention: Village's functional complex that strenghens cohesion of clan

未走几步，到了一个地方的门口，又像破庙的大门一样，然而这里却很狭，破庙的门却要宽许多。路是不平的，两边堆着碎石残瓦。不到四五步便走进去了。

天井中没有石板，是泥地，走上石阶约十余步便是神龛。神龛中放着神主（约有三隔，中间放着始祖的神主，但现在记不清楚了），外面嵌满了玻璃，玻璃窗上已生满了尘埃，中间的玻璃也有碎的了。神龛面前放了一张破烂的桌子，石阶两旁各有一排栏杆，上面有几扇窗户，但现在已没有了。靠着右边墙壁走过去有一道小门，四伯祖把门拉开，我们走去原来是一片堆着碎瓦的地。屋顶是漏的，抬头可见着青天。靠着栏杆放着几块破砖，围成一个小炉子，上面放着一个大瓦罐，是盖着的，不知里面煮的是什么东西。天井中放一张桌子，一个成衣匠在那里缝衣。这就是我们的祠堂！

——巴金·《塘汇李家祠堂》

形式的变化潜能一：
祠堂功能的竖向发展

Formal Transformation Potential I:
Vertical development of ancetral hall

形式的变化潜能二：
祠堂在乡村肌理中的重新置入

Formal Transformation Potential II:
Refill of ancestral hall into the current rural texture

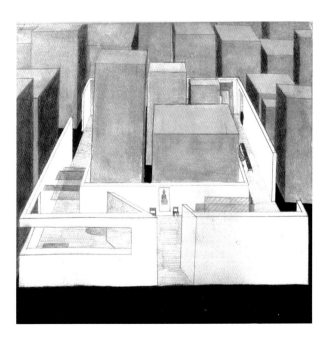 

[戏台]

　　古溪村赵氏祠堂内有一个完整的戏台。大部分农村地区的戏台都在祠堂之中，也有个别地方的戏台就是在村中某个空地上，或者穷一些的地方就在有戏班子来的时候临时搭台搭棚。古溪村比河阳更深入腹地，当然不如河阳兴旺，但是拥有自己的戏台却足以令村民自豪。观看社戏是日常乡村生活的高潮，而能在本村看戏更是一个宗族乃至一个村子的脸面。

　　现在的戏台下青草悠悠，显然是许久没有村民围坐了。但时有顽童在戏台上的"出将"和"入相"的门中窜进窜出，打打闹闹。一个不知从哪里找来戏中大将的背靠要将起来，一个则躲在台子下面。

　　不过一到年节边戏班子到来的时候，这里就会是完全不同的光景了。即使现在传媒发达，但是做戏文的热闹村民们是一定要去凑的，这也是小孩子们最为盼望的活动了。鲁迅在《社戏》里的孩子们偷偷划船去看社戏的描写至今让人印象深刻。沉沉的黑夜中的远处戏台上的灯火，好像是充满着光晕的梦幻仙境。

[Village Stage]

There is a well reserved old stage in Zhao's ancestral hall in Guxi Village. Most of the stages in rural areas are set in the ancestry halls. On very few occasions, the stage is built upon some vacant land in the village. In some poor areas, only temporary stage or shed is afforded to be set up when a troupe comes. Guxi village is located in a hinterland deeper than Heyang. Although there is no comparison of prosperity can be drawn between the two villages, villagers in Guxi village are proud enough to have their own stage. Watching theatrical performance is the climax of daily rural life. Being able to watch it in one's own village brings the indignity of a clan or even the whole village to another level.

The grass beneath the stage is luxuriant. It is evident that it has been long time since the gathering for the last performance. Nevertheless, kids are often seen jumping in and out of the backstage doors on the stage playing dramatic characters, or playing peekaboo hiding under the stage.

Once a drama troupe arrives during the Spring Festival, it would be a completely different scene here. Although the mass media has been advanced today, no one in the village would like to miss the fun the drama would bring once a year. This is also some activity kids have been expected all year long. In Lu Xun's fiction ''Village Drama'', the description of the country kids sneaking to see the drama by boat is still impressive. The lights on the distant stage in the deep night seemed like a dreamland full of halos.

原型的心灵意向：村中心做戏的地方 —— 儿童心中的有着光晕的梦幻仙境

Archetypal Intention: Performing stage at the center of village - a wonderland with halo in children's mind

两岸的豆麦和河底的水草所发散出来的清香，夹杂在水气中扑面的吹来；月色便朦胧在这水气里。淡黑的起伏的连山，仿佛是踊跃的铁的兽脊似的，都远远地向船尾跑去了，但我却还以为船慢。他们换了四回手，渐望见依稀的赵庄，而且似乎听到歌吹了，还有几点火，料想便是戏台，但或者也许是渔火。

……

最惹眼的是屹立在庄外临河的空地上的一座戏台，模糊在远处的月夜中，和空间几乎分不出界限，我疑心画上见过的仙境，就在这里出现了。这时船走得更快，不多时，在台上显出人物来，红红绿绿的动，近台的河里一望乌黑的是看戏的人家的船篷。

『近台没有什么空了，我们远远地看罢。』阿发说。

这时船慢了，不久就到，果然近不得台旁，大家只能下了篙，比那正对戏台的神棚还要远。其实我们这白篷的航船，本也不愿意和乌篷的船在一处，而况并没有空地呢……

在停船的匆忙中，看见台上有一个黑的长胡子的背上插着四张旗，捏着长枪，和一群赤膊的人正打仗。双喜说，那就是有名的铁头老生，能连翻八十四个筋斗，他日里亲自数过的。

……

我不喝水，支撑着仍然看，也说不出见了些什么，只觉得戏子的脸都渐渐的有些稀奇了，那五官渐不甚分明，似乎融成一片的再没有什么高低。年纪小的几个多打呵欠了，大的也各管自己谈话。忽而一个红衫的小丑被绑在台柱子上，给一个花白胡子的用马鞭打起来了，大家才又振作精神的笑着看。在这一夜里，我以为这实在要算是最好的一折。

……

月还没有落，仿佛看戏也并不很久似的，而一离赵庄，月光又显得格外的皎洁。回望戏台在灯火光中，却又如初来未到时候一般，又漂渺得像一座仙山楼阁，满被红霞罩着了。吹到耳边来的又是横笛，很悠扬，我疑心老旦已经进去了，但也不好意思说再回去看。

——鲁迅·《社戏》

165

形式的变化潜能一：戏剧和事件在不同场地的框景
Formal Transformation potential I: Framed stage of dramas and events on varied sites

形式的变化潜能二：观与被观的空间关系
Formal Transformation Potential II: Spatial relationship between the spectating and the spectated

形

胜

目

[月塘]

　　河阳一带有新建溪从山上潺潺而下，周边的山溪也汇聚到大溪之中，这里属于上游，山泉清澈。流经的村子都充分利用这里的水系，将水引入村中沟渠和池塘，以便日常取汲饮水和救火之需。水塘也是乡村风水的重要组成部分，旧时人们以水为财，在合适的位置引水入塘，代表着聚宝敛财。不仅是房前屋后有半亩方塘，村口或是村中心一般也有大一些的中心水塘。韩畈村在村口有个极大的塘，而河阳村的大塘则是在村子深处。村子的大水塘经常呈新月形或者半月形，取"月满则亏，水满则溢"之意，体现留有余地的传统处世哲学。西岸村的中心月塘像是一弯窄窄的新月，古溪村的中心月塘是饱满的半月。

　　不管是白天还是黄昏，粉墙黛瓦所包围的这一围月塘是村子里最美的景色。白天这里是"天光云影共徘徊"，村子里大大小小的水塘是一面面锃亮的明镜，远处如黛的青山和近处斑驳的墙影在镜中的影像被空中翻卷的白云不断变幻着。夜晚这里是"月静池塘桐叶影"，水中的月影偶尔会被鱼儿浮上来的涟漪打乱。这半亩月塘虽小，却是归隐田园的文人心境的写照，空灵无为而载宇宙万千变幻。

　　不管冬日还是夏夜，月塘也必定是村子里最有人气的所在。冬天，老人们在开阔的水塘边晒太阳。夏日的傍晚这里是纳凉的天地，大人在竹躺椅上摇着蒲扇，小孩们就在支起来的竹床上玩耍。

[Moon Pond]

　　The Xinjian Stream flows through the area of Heyang, joined by the creeks down from adjacent mountains. It is the upstream here with crystal clear water. The villages along the stream make full use of the water, drawing it into the ditches and ponds to satisfy the daily needs of freshwater and fire fighting. Ponds are also an important part of rural Feng Shui. People believed water represents wealth. Making a pond in an appropriate location and drawing water into it could accumulate wealth. So ponds were made wherever possible, close to a house, or right in the center of a village. There is a very large pond at the entrance to the Hanfan Village, while the big pond of Heyang is deep inside of the village. The big pond of a village is often crescent or half-moon shaped, taking the meaning of "the moon waxes only to wane, water brims only to overflow", which reflects the traditional philosophy of life. The center pond of Xi'an village is shaped like a narrow crescent, and the one of Guxi village is like a half moon.

　　Whether it is during the day or at night, the moon pond surrounded by white walls and black tile roofs is the most beautiful scene of the village. During the day, it is "where sunlight and clouds linger and leave" (Zhu Xi's poem "the Book"). The ponds scattered in the village are like bright mirrors. The dark green mountains in the distance and the nearby mottled walls are mirrored in the water, while the image is constantly disturbed by the reflection of the rolling clouds. At night it is "a quiet pond with the shadow of the moon and parasol leaves". The reflection of the moon on the water surface is sometimes blurred by the ripples made by swimming fish. Although this half acre moon pond is small, it is an implication of the philosophy of those scholars who chose to live in the countryside. Crystal clear mind carries the whole universe.

　　Whether it is winter day or summer night, the moon pond must be the most popular place in the village. In winter, the elderly gather around the pond basking in the sun. At summer night, it is a place to enjoy the cool, where the adults lie in bamboo lounge chairs with fans at hands and the children play on bamboo beds.

原型的心灵意向：被村宅包围的半亩水塘映射着变幻无穷的天光云影，也映射出乡村文人的归田心境

Archetupal Intention: A small pond embedded within village villas reflects ever-changing sky and clouds, as well as rural scholars' heart-mind

月光如流水一般，静静地泻在这一片叶子和花上。薄薄的青雾浮起在荷塘里。叶子和花仿佛在牛乳中洗过一样；又像笼着轻纱的梦。虽然是满月，天上却有一层淡淡的云，所以不能朗照；但我以为这恰是到了好处——酣眠固不可少，小睡也别有风味的。月光是隔了树照过来的，高处丛生的灌木，落下参差的斑驳的黑影，峭楞楞如鬼一般；弯弯的杨柳的稀疏的倩影，却又像是画在荷叶上。塘中的月色并不均匀；但光与影有着和谐的旋律，如梵婀玲上奏着的名曲。

荷塘的四面，远远近近，高高低低都是树，而杨柳最多。这些树将一片荷塘重重围住；只在小路一旁，漏着几段空隙，像是特为月光留下的。树色一例是阴阴的，乍看像一团烟雾；但杨柳的丰姿，便在烟雾里也辨得出。树梢上隐隐约约的是一带远山，只有些大意罢了。树缝里也漏着一两点路灯光，没精打采的，是渴睡人的眼。这时候最热闹的，要数树上的蝉声与水里的蛙声；但热闹是它们的，我什么也没有。

——朱自清·《荷塘月色》

形式的变化潜能一：儿童的月塘

Formal Transformation Potential I: Moon pond in children's imagination

形式的变化潜能二：城市围合中的月塘梦境
Formal Transformation Potential II: Moon pond embeded within urban environment

[锄云]

往河阳村的深处走，里面有一个很大的水塘，在水塘的西北一侧先是会看到一个十八间合院。合院正面两侧的山墙是跌落三阶的马头墙，中间夹着的正门却不在中间，而是开在左侧的山墙所对的厢房通道上。门洞和上面的门罩之间有写着"锄云"两字的门额，含义令人费解。

合院边上是一个十八间的"日"字形大宅，大宅面向水塘的厢房外墙的外面还有一层高的围墙，斜斜地过来和前面的合院外墙连为一气。围墙的上方能看到后面外墙上厚重的砖雕门罩和窗罩，高高低低的。大宅的方向和合院相比转了一个直角，每一片山墙都是垂直于水塘的，上面是跌落五阶的马头墙。这里的马头墙涩檐深，戗角起翘曲高，真的如黑马仰头长啸。

坐在水塘边，斜斜地看向这片高低错落、重重**叠叠**的马头墙群，天空中白云飘过，霎时明白了"锄云"的意思。这些马头墙就像农家的锄头，将天上的白云锄断。

[Hoeing Clouds]

Going deeper into the Heyang village, there is a large pond inside. On the northwest side of the pond, there is an Eighteen-room house. The gables on both sides of the courtyard are three-step styled horse-head walls. The main door is not in the middle as usual, but opens on the left side of the gable on the passage to the wing rooms. Between the doorway and the decorative cornice above, there is a plaque inscripted with two Chinese characters ''Chu Yun'' (Hoeing Clouds), the implication of which seems puzzling.

Beside this mansion is another Eighteen-room house. The exterior walls of the two houses are connected into a long wall facing the pond, with various decorative brick cornices above the doors and windows. The orientation of this big house makes a right angle to its neighbor, which makes its gables perpendicular to the pond. This house has five-step styled horse-head gables with eaves deep and curves risen high, which is really like a dark horse with head up roaring.

Sitting by the pond, looking at this group of horse-head walls scattered and overlapped, and white clouds floating in the sky, I suddenly understood the meaning of ''Hoeing Clouds''. These horse-head walls are like a farmer's hoe, clipping the white clouds in the sky.

原型的心灵意向：具有传统江南意象的乡村天际线建构

Archetypal Intention: Constructing village's skyline of traditional image of Jiangnan area

早起的阳光，越过高耸的马头墙，倔立的瓦松，沐浴着生命的礼赞。

鸟雀唧啾，在屋檐上跳来跳去，唤醒了驳岸边的杨柳。

挎着布袋书包的小学生，在廊棚里追逐嬉闹，脚步匆乱了古镇的晨光。

香花桥上露珠尚未消褪，湿漉漉的向着悠长的香花弄蔓延。

数着步子，走过结着青苔的石板路，保圣寺的晨钟响起。

紧随着佛国梵音，一墙之隔的小学堂，也即将拉响上课的电铃声。

一位清瘦的年轻人，穿着青灰色长袍，捧着一摞书，立于校门外，三三两两的小学生从他身边经过，恭恭敬敬鞠了一躬，喊道，『先生早！』

年轻人微微躬身，笑着回应，『早。』

那声『早』，是标准的苏州口音，说的认真且又软糯。

——叶圣陶·《破冬的一缕春风》

老屋终于在一场大雪中轰然倒塌了。

当我来到坍圮的老屋面前时，老屋已是一片狼藉。惟有雄伟苍老的马头墙，巍然耸立在山墙两边，犹如两头异常勇猛的困兽，依然保持着往昔桀骜不驯的雄风。

然而，它的脊檩，它的后进，已是断壁残垣，惨不忍睹。

——史良高·《最后的老屋》

形式的变化潜能一：
戗角的拟人化

Formal Transformation Potential I:
Anthropomorphization of corbel gables

形式的变化潜能二：
马头墙的负形

Formal Transformation Potential II:
Neqtive form of horse-head walls

形式的变化潜能三：
马头墙的构成

Formal Transformation Potential III:
Composition of horse-head walls

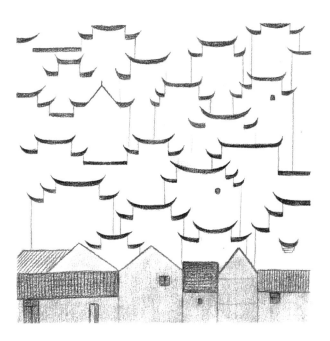

[路亭]

亭者，停也。

在乡村地区，从比较偏远的村子到集镇是要赶不少路的，尤其是那些深山之中的村子。有些偏僻的地方，甚至途中都不能碰到一个有人烟的地方，即所谓前不着村后不着店。这个时候，如能有个亭子暂且休息一下，那是极好的。乡野之中，亭子的样式无关紧要，只要一小片茅草顶，就宣示了一小片属于人的领地，而不至于让人觉得在乡野之中惶恐无助。亭子虽小，却是占据广袤的村野之上的重要的人造物。

河阳一带是山谷之间的狭长平原，从集镇一路到深山之中有不少的脚程。这一方百姓一直将建路亭视为善举，方便他人，也方便自己。离集镇最近的韩畈村的村口水塘就有一个新建的六角亭，曰"聚心亭"。走到上面的村子，古溪村口也有一个一模一样的六角亭，曰"连心亭"。路上还时不时地在公路两侧看到一些非常简陋的棚子，是农民暂存农具和耕作时休息的地方，当然以前交通不方便的时候，路人也可以随时进去小坐。这样的棚子由村民随兴而建，实在谈不上什么章法，但是倒也符合《园冶》中关于亭子"随意合宜则制，惟地图可略式也"的说法。

兰溪曾经有一个亭子，因明末李渔取名为"且停亭"而闻名。且停，且停，且在这亭子里停一停，歇歇脚，怎么不能叫且停亭？这又何尝不是一种归园田居的生活态度？

[Pavilion]

A pavilion, is a stop. (They share the same root in Chinese characters.)

In rural areas, people had to travel a long distance from their remote villages to the market town, especially those located deep in the mountains. When travelling through some remote places, it could be difficult to encounter even one single person. That is called in the middle of nowhere. At this moment, a pavilion serves a best purpose of rest. In the countryside, the style of the pavilion is not important. A small piece of thatched roof claims a small piece of territory which belongs to human and helps human fight against the helpless feeling in the wild. As small as a pavilion is, it is an important man-made object that dominates the vast rural wilderness.

It is a narrow plain between valleys in Heyang area. There is a long distance from the market town of Xinjian to the deep mountains. So building a pavilion is regarded as a charity that provides convenience for everybody. There is a new hexagonal pavilion nameed ''Juxin Pavilion'' (Gathering Hearts) at the entrance of Hanfan Village. Walking up to Guxi Village, there is an identical pavilion, named ''Lianxin Pavilion'' (Connecting Hearts). Along the road, there are simple sheds which are used as temporary storage of farming tools and farmers' rest place. When the transportation was not convenient in the past, anyone could go inside for rest. Such sheds were built by farmers at their free will with no construction rules at all. However, it echoes the description about the pavilion in the masterpiece ''The Craft of Gardens'', which says, ''a pavilion can be built at will, if only it fits the environment.''

There was a famous pavilion in Lanxi once, well known for the name of ''Please Stop Pavilion'' given by Li Yu of the late Ming Dynasty. Stop, stop, please take a stop and have some rest in this pavilion. Isn't this some life philosophy of returning to the countryside?

原型的心灵意向：乡野中任意一个顶盖界定的临时歇脚之地

Archetypal Intention: A rest place defined by a canopy on the road for farmers or travellers

赶市的村夫农妇，或者担着辛苦经营的菜豆瓜果、鱼米柴草到街上求售，或者提篮挑筐，到街上去买办日用杂物、农事工具，或者因为借贷无门，挟些不值钱的衣物破烂上当铺去质钱……每天清早，朝阳初窥田野，便沐晨风，带晓雾，从村里出发，哼哼唧唧，形成行列，快步赶上镇去，直要到事毕功成，事倍功半，或者事败功亏，才循原路赶回村里。

奔波忙碌了半天，人是倦了；而『不如意事常八九』，乘兴而去，常常败兴而归。心情懊丧，双脚沉重，生理和心理的倦怠形成双倍的压力。幸而半路有个路亭，排闷迎人，容他们且住为佳，使身心暂时有个着落。吹一阵凉风，扯一阵闲话；再闲闲地抽一筒早烟，让生命获得片时的苏息，好再鼓起勇气，继续上路。不巧遇上意外的天气变幻，更可以在路亭里求得荫庇，聊避风雨。

试想这对疲倦的旅人，是何等温煦的抚慰！

路亭所处的位置，不但富于实用价值，又多似高明的画师布局，引人入胜。有的点缀田畴广野中间，『前不把村，后不着店』，亭亭玉立，不但使无根的流浪人，无处投宿时借此歇夜，对田头劳作的农民，这又是天然的接待天涯沦落的平原减少单调之感，还便于旅途修长的过客及时小驻，更可以地，日中时刻，可以静坐进餐，夏避炎阳。有的高踞岭背，峰回路转、两村交界之处，翼然一亭，挺秀如画。山行较平地费力，行人跑到岭上，大都气息咻咻，汗流浃背，在路亭的石条凳上坐憩片刻，听山风苏苏从树梢掠过，便利行人投下一身清凉。有的筑在河滨，面临盈盈的流水，傍着霭霭的绿荫，随意歇脚，等待摆渡或过往的船只。……

离我老家不远，有两个路亭，是我幼年踪迹最频之处，得闲还常去盘桓。大江沿有个过渡亭，好像建筑得特别讲究，地位大，墙壁石凳，整齐可观，临河还有宽广的双面『埠道』；一到夏季，晚霞掩映中，那里差不多成了公共浴场。亭前石柱上，刻着两副对联，记得其中的一副是：山色湖光，四时佳兴；早南晚北，廿里官塘。

对联虽然并不高明，但山色湖光，并非虚语。普通路亭，虽也有对联点缀，却无非是『稍安毋躁』、『小坐何妨』之类，这样『风雅』的对联是例外。不过疲倦的行人，谁也不计较这些。

——柯灵·《路亭》

形式的变化潜能一：
路亭的形式变异

Formal Transformation Potential I:
Transmutation of pavilions

形式的变化潜能二：
不同环境中的路亭

Formal Transformation Potential II:
Pavilions in different environmental setting

[桥]

　　有大溪流过，自然这一带也有各式的桥。

　　大溪流到河阳村口，河道已经渐宽，其上有一座清代所建的五孔石桥 —— 公济大桥。据《义阳朱氏家谱·公济桥记》，"肇始于道光庚戌，辟洞五空，长三十二丈，广一丈五尺，越咸丰癸丑桥成。"近看石桥厚重，石桥每个桥墩都带分水金刚雁翅。退远到半里外村妇洗衣的河埠头回望，在古树繁茂的枝条后面，厚重的桥身中五个大桥洞和水中倒影相合，构成五轮圆月，托起上面平直的桥线，又显得无比清秀和安静。

　　溯大溪而上，交雅村口也有圆拱古桥一座，曰福荫桥，据说有三百多年历史。古桥掩映在周围的大树之下，桥下麻鸭一片。

　　再往上走到马堰村，村子上面就是白马水库，水库大坝和村子落差很大，坝下的水道中大石堆叠，水库所积蓄的溪水就从这些大石间奔泻而下到村边的河道之中。这里已然是山水之间，在村边小路上闲逛，突然看到一座无名铁桥横跨在草木丛生的溪谷中，简单的人字桁架，锈迹斑斑显得有些年头了。铁桥没有什么古意，但是尺度宜人，在这山水之间，又有一番意境。所谓山水是在人的观照之后由物象到心象的转化，所以山水是有人迹的。有桥就有人迹，人与自然在此神遇。

[Bridge]

There is a big stream flowing through the area, and certainly there are bridges of different kinds.

When the big stream flows to the Heyang village, the streamway gradually becomes wider, above which there is a five-arch stone bridge built in the Qing Dynasty - Gongji Bridge. According to the "Family Tree of House of Zhu at Yi Yang - Gongji Bridge Record", "the construction of the bridge originally started in the year of Gengxu in the Daoguang period. It was about thirty-two zhang (1zhang=3.3meter) long and one zhang and a half wide with five arches. It was finished in the year of Kuichou in the Xianfeng period." Take a closer look, the stone bridge is heavy, each pier of which is equipped with a water diversion wing. Retreat a half mile away and stand on a laundry quay looking back, behind the lush branches of the ancient trees, the five arches are joined by their reflections in the water to constitute five full moons holding up the straight line of the bridge, making a quiet and elegant composition.

Walking up along the big stream, there is also an ancient bridge with round arch at the mouth of Jiaoya village, called Fuyin Bridge, which is said to be more than three hundred years old. The ancient bridge is hidden under the surrounding trees. Sometimes a flock of ducks is seen floating down under the bridge.

Further up, there is the village of Mayan. Above the village is the Baima Reservoir. There's large drop between the reservoir dam and the village. Hugh boulders stack in the waterway below the dam. The reservoir water cascades through these boulders from time to time down to the creek in the village. This is already natural landscape. An unnamed iron bridge is there spanning a valley of vegetation, with simple flat truss that appears to be rusty and aged. This iron bridge has nothing to do with antiquity, its scale is pleasant though, which creates a special kind of interaction with the landscape. Landscape in Chinese, or Shanshui, is truly a conversion from physical image to heart image after contemplation. So human presence is important in Chinese Shanshui. A bridge implies human presence. This is where the spiritual encounter of human and nature takes place.

原型的心灵意向：乡村的交通联系，也是和山水意境的精神联系

Archetypal Intention: Physical connection in the landscape as well as spiritual connection with the landscape

他们三个人今天一齐游八丈亭。小林做小孩子的时候，时常同着他的小朋友上八丈亭玩，琴子细竹是第一次了。从史家庄这一条路来，小林也未曾走过，沿河坝走，快到八丈亭，要过一架木桥。这个东西，在他的记忆里是渡不过的，而且是一个奇迹，一记起它来，也记起他自己的畏缩的影子，永远站在桥的这一边。因为既是木架的桥，又长，又狭，又颇高，没有攀手的地方，小孩子喜欢跑来看，跑到了又站住，站在桥头，四顾而返。实际上这十年以内发了几次山洪，桥冲坍了重新修造了两回。依然是当初的形式。今天动身出来，他却没有想到这个桥，坝上都是树，看见了这个桥，桥已经在他的面前。他立刻也就认识了。很容易的过得去，他相信。当然，只要再一开步。他逡巡着，望着对岸。细竹请他走，因为他走在先。他笑道：

『你们两人先走，我站在这里看你们过桥。』

推让起来反而不好，琴子笑着首先走上去了。走到中间，细竹掉转头来，看他还站在那里，嚷道：

『你这个人真奇怪，还站在那里看什么呢？』

说着她站住了。

实在他自己也不知道站在那里看什么。过去的灵魂愈望愈渺茫，当前的两幅后影也随着带远了。很像一个梦境。颜色还是桥上的颜色。细竹一回头，非常之惊异于这一面了，『桥下水流呜咽，』仿佛立刻听见水响，望她而一笑。从此这个桥就以中间为彼岸，细竹在那里站住了，永瞻风采，一空倚傍。

这一下的印象真是深。

过了桥，站在一棵树底下，回头看一看，这一下子又非同小可，望见对岸一棵树，树顶上也还有一个鸟窠，简直是二十年前的样子，『程小林』站在这边望它想攀上去！于是他开口道：

『这个桥我并没有过。』

说得有一点伤感。

『那一棵树还是同我隔了这一个桥。』

接着把儿时这段事实告诉她们听。

『我的灵魂还永远是站在这一个地方，——看你们过桥。』

是忽然超度到那一岸去了。

——废名·《桥》

形式的变化潜能一：
桥的变异

Formal Transformation Potential I:
Transmutation of bridge

形式的变化潜能二：
桥和山水的构成

Formal Transformation Potential II:
Composition of bridges and landscape

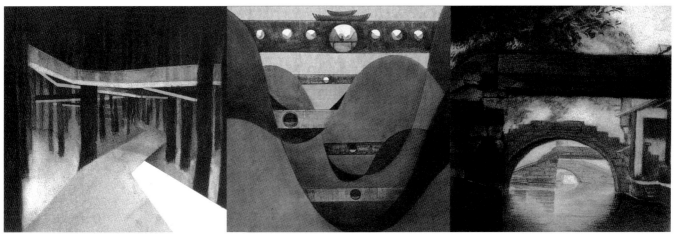

第四章

视角与重构

邻里关系
乡村经济
乡村居住
宗族社会
自发建造
乡野景观
......

沿着河阳大溪的七个村子，除了河阳是国家级重点文物保护单位使得它的问题基本只是单纯的历史保护之外，其他村子则面临着这些中国乡村都面临的共同的现实问题。

Neighborhood relations
Rural Economy
Vernacular Dwelling
Patriarchal Society
Spontaneous Construction
Country Landscape

...

Among the seven villages along the big stream of Heyang, excpet Heyang which has a focus issue of preservation as a major historical and cultural site protected at the national level, while the rest are faced with these same practical issues encountered by all other Chinese villages.

[农妇的曲水流觞]

视角 ------------------------ 邻里关系
村落 ------------------------ 韩畈
原型 ------------------------ 浣渠、道坛、月塘、古树

　　和城市一样，乡村的邻里关系也在淡化。
　　乡村的邻里关系某种程度上是由农妇主导的，一起洗衣，一起家长里短。韩畈村有水渠环绕，在沟渠边洗衣洗菜的农妇们一丝不苟地工作着，但当聊到开心的话题时，农妇们和曲水流觞的文人们在兴致上似乎是没什么区别的。
　　门口浣渠上的青石板就是农妇们的领地，以这个领地为核心重新组织乡村户外生活空间为保护和发展乡村邻里关系提供一种新的可能途径。

村井

鱼沼

堂池

浣塘

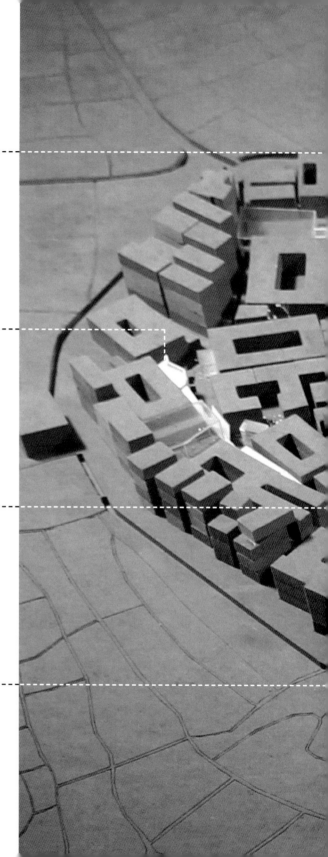

[番薯生产合作社]

视角 ----------------------- 乡村经济
村落 ----------------------- 岩山下
原型 ----------------------- 簟基、道坛、村巷、古树、墙垣

　　对于村民而言，经济收入自然是排在首位的大事。这一带的村民有制作和销售番薯干的传统，但是一家一户毕竟能力有限，如果能集合起来，善于种植的种植，善于找供销门路的去销售，为农产品增添更多的附加值，对于村民而言是没有任何坏处的。

　　在农村不管是老房子还是新建房，几乎都有专门堆放农具和存储农作物的房间。如果村民愿意尝试将这些住宅当中和劳动生产相关的面积贡献出来，形成一个共享的集体生产合作的空间，这将形成一种全新的乡村居住模式。

　　村民们在番薯收获季节时支起来的丝网棚是这个番薯生产合作社的空间结构的灵感来源。

斜撑的簟子

场所意象

概念模型

概念模型

独立起居　独立起居　独立起居　独立起居　独立起居

连通庭院

集体生产合作社通道　　　　红薯生产合作社间

私密露台　　集体晒场

独立起居

独立起居　　连通庭院　　集体仓库

211

[月影中的十八间]

视角 ———————————— 乡村居住
村落 ———————————— 西岸
原型 ———————————— 映月、道坛、炊烟、墙垣、户牖

　　西岸村有个自建楼，七八年了，一直没有完工的迹象，建筑立面上不断重复着月洞门的意象要素。农民对于中国传统要素有着某种不自觉的向往，只是对于形式的操弄不甚熟练。
　　十八间又是这一带传统的民居形式，围绕中间的庭院大致有十八个房间，可以容纳一大家子。但是当下农村的家庭结构早已小规模化，在十八间这样的传统民居的传承方面需要充分考虑如何适应新的家庭结构，如何承载曾经的文化意象。

场所意向

平面肌理

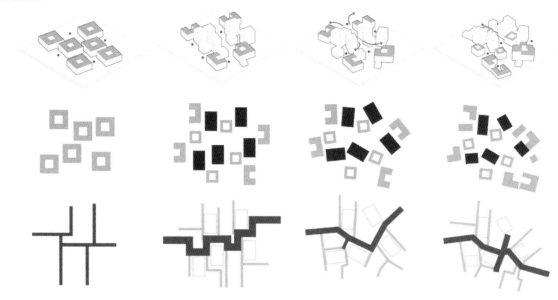

墙与月影

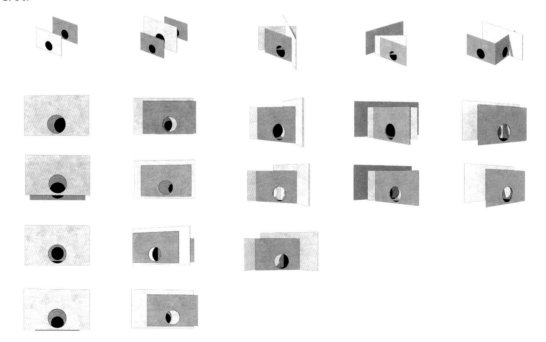

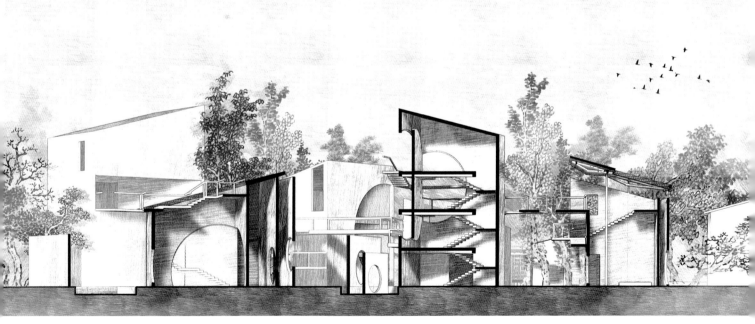

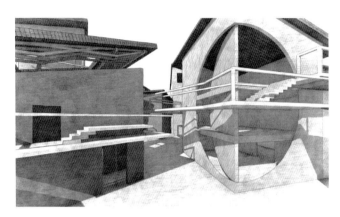

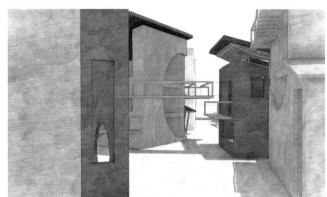

[留守儿童的天堂]

视角 --------------------------- 宗族社会
村落 --------------------------- 古溪
原型 --------------------------- 月塘、戏台、宗祠、路亭、古树

　　城镇化深度地瓦解了乡村的宗族关系，青壮年离开乡村，留守在乡村的是儿童和老人。但无拘无束的乡村如同鲁迅笔下的百草园，也是儿童的天堂。
　　月塘中的云影漂浮到村子各处，将学校、戏台、路亭、山坡、古树转译成留守儿童的隐秘乐园。

山亭

叠石

水榭

折桥

场所意向

乡村儿童中心

新水塘

老水塘

[乡村建筑师日记]

视角 ----------------- 自发建造
村落 ----------------- 交雅
原型 ----------------- 墙垣、户牖、道坛、炊烟、锄云

　　当代中国乡村最为人诟病的就是农民房的美感缺失。但是这不能归罪于农民，缺乏可参照的样本，他们只能以自己的理解，用打工时学到的建房技巧和市场上能买到的建筑材料完成农民房的自建。
　　农民是不缺创造力和动手精神的，如果他们能得到一些指引，会达成什么样的自建？三位没有太多设计经验的低年级建筑专业学生代入三位农民建筑师的身份当中，从不同的类型研究入手，通过经验总结，得出自己想要的样式。

场所意向

部件类型研究

檐下

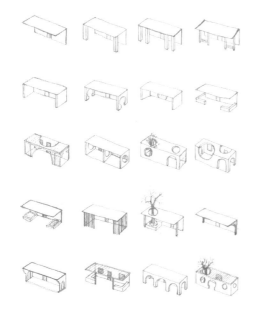

晒台

材质类型研究

墙垣

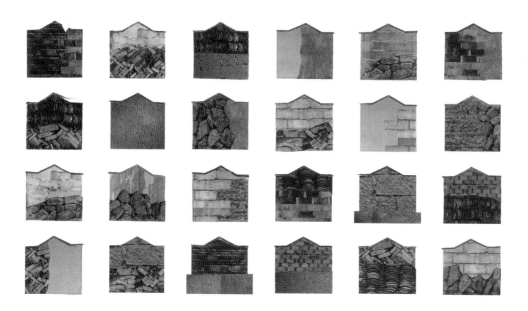

2015/05/11			
当日工程施工部位	场地	当日工程施工内容	浇制首盘地基 第二盘地基支模
浇制首盘地基			
第二盘地基支模			
施工安全场地情况	安全	施工安全人员情况	良好

2015/10/25			
当日工程施工部位	主体封顶	当日工程施工内容	做铝合金窗
主体封顶			
做铝合金窗	费用：选了广东的某品牌铝材，1.0mm厚度。28元/公斤，大小总共55个窗（含2、3楼各一推拉门）。用料288公斤。铝合金工上门拼装。工费40元一平方，包窗框灌水泥浆，包磨页玻璃安装。包抹窗安装。总价约11000元。		
施工安全场地情况	安全	施工安全人员情况	良好

2015/11/26			
当日工程施工部位	主体封顶	当日工程施工内容	铝合金门窗安装完成
插曲二 "客厅左侧的书房侧窗，原窗宽2.3米，盖好后发现与客房前面的窗或者平窗了，铝合金师傅建议我把窗小更好些。我采纳了建议，把工头叫回，加砌砖块，窗框缩小1.8米，缩小到1.8米的倒窗，上下也材称，原来凸出的窗线也，工头说等林外墙水泥未时凹切割掉，但我觉得留着也不错，将错就错。我决定就保留这个造型。" 铝合金门窗安装完成 买木料 "我父亲做木的好兄弟林师傅前来工地考察，老父说让林师傅带我去了解一下木村本地市里的木料卖得比较贵，说邻近的县城就能便宜不少。我打算楼立柱及扶手，大门及房间门都用实木，林师傅说有3方应该够。放要格买不起，黄楷木的6000一立方，便是可以考虑。但林师傅说黑果不做那些实木的东西，"龙凤廉"也不错，现在也比较多人喜欢用，木色及纹理都好看，也不变形。还便宜2000多一方。最后定了要"龙凤廉"原木，林师傅选了3根，计有6个立方。"			
施工安全场地情况	安全	施工安全人员情况	良好

三月十六号

问了问老板，拆梁或施木桁架都是可以的。三层阁楼有点矮，做把桁架嵌到墙里，然后再开窗。二层的露台也要等一个屋顶，可以也架个披屋顶，跟着子的披屋顶斜一下，不接在一起，还能再向外挑出一个小阳台。这个二层的屋架其实还能不跟一层的结构连在一起，省点料子。

其实还是把屋顶的结构和露盖出来，把木头打磨上漆。麻烦是麻烦，不过房子人家做的真的是挺好看。

今天还看到对门在搭露台上的阳光房了，他们盖得有点快，这会都要盖完了。木支撑全都露出来，算一层和克板，怪挺好看的，当时一层他们做的，也没用撑甲，搬起手行，做起来还挺快的，平天就搭完了。过去看了看他们怎么做的，他还接拉了我一份图，能看出来是一层叠一层的上去的。

四月四号

这两天屋架就要搭好了，放整围墙了。虽然村子的老房子都是泥土，但是其实不是很想揭布土，土和模具都得现把规规搭，费事不说之后还容易坏，还是用围墙砖砌墙，越用青砖，看起来就挺好看的。

做空斗省点料子，一眼一丁，把砖斗子放外面。砖墙中间就点碎砖，不过就脚也是得用扁瓦。是想要比较省的效果，村墙墙得火啊都要做简单点，不过简单的做法也要用两三层普通青砖做叠浇出纳，不这样又会跟老房子像像了。

抹灰的话，要不还是做清水墙吧，也不用勾缝抹缝处，做点砖墙的镶窗，就墙做点花纹啊抹搭捆啊，这样应该挺挺好。最好再在面种点树，种梅树？老人应该会挺喜欢的。

四月五号

今天看施工队施工，砖墙的砖宽比我想得要，我是知道砖砌是一定要用丁顺梳合的扁砌的，砌五到九层，必须要皇单数然后七层层要多，上身墙也要比脚脚处差，而且知道一个省材的法子，脚脚也一定要满面至灰妈随砌，只身墙就可以在四面用瓦刀挂灰，哎，挺炸，不过这点是想要平整点的墙，师傅跟我说上身墙脚也可以不灰。

这时候拿着铺屋瓦也快了，抬的瓦片才挂了一点，这两大一扎是点把瓦忘了，明天拿再去看看剩一点有，不过应该不能再有到一点。得要早上，到晚了没有了，能省就省，而且最后还挂好着的。

[二十四桥明月夜]

视角 ------------------------ 乡野景观
村落 ------------------------ 马堰
原型 ------------------------ 桥、锄云、月塘、古树、路亭、戏台

  乡村生活的意趣在于人能在此与自然神遇，乡野中的一座座桥就是最好的见证。马堰村在上流溪谷就有一座无名铁桥，虽锈迹斑斑却让这处偏僻乡野瞬时灵动起来。
  在这个山环水抱的小山村，以重重叠叠的桥为针线重新编织这片乡野的肌理，展现二十四桥明月夜的当代意境。

场所意向

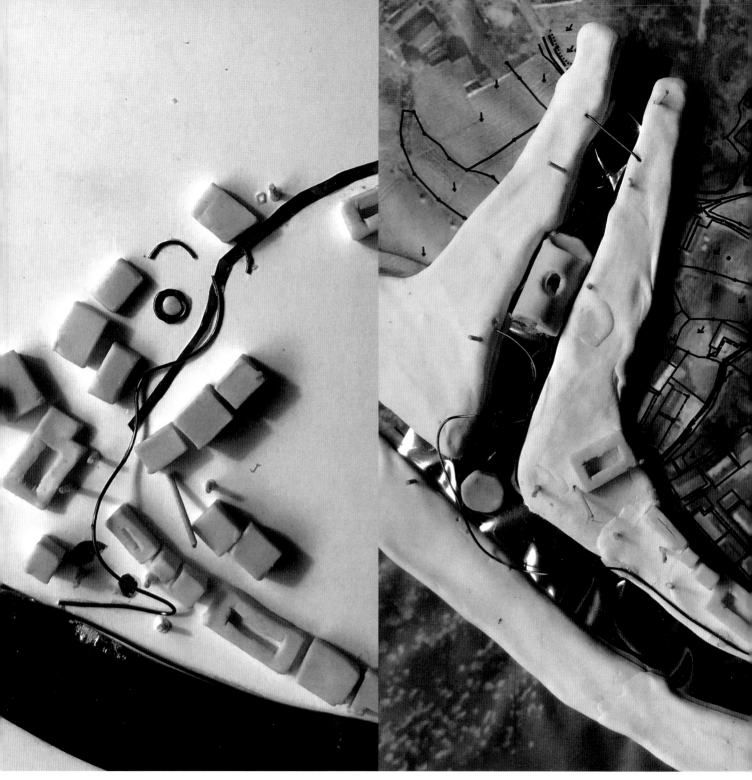

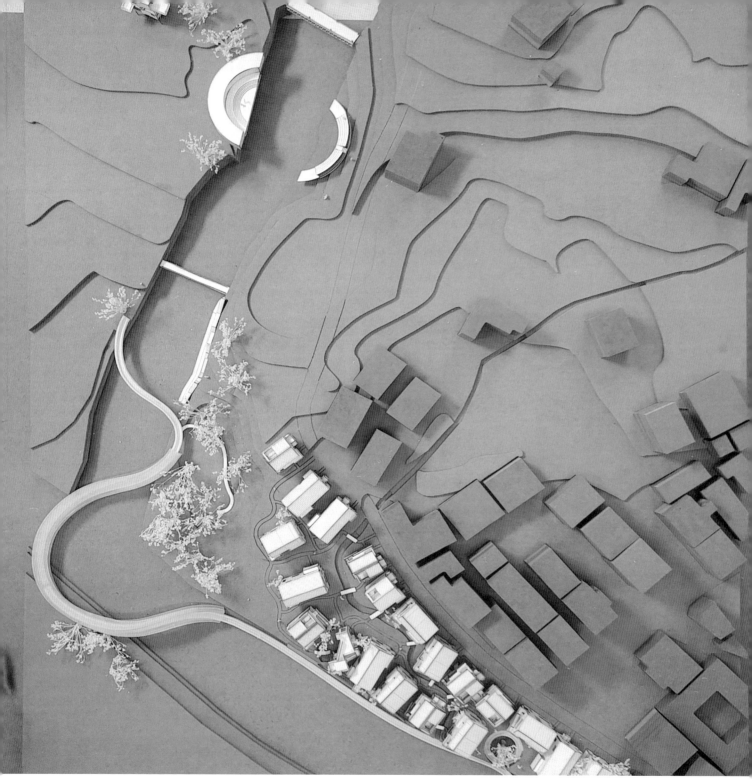

木编拱桥

钢桁架桥

235

[参与名单]

本书的形成过程也是一次次教学实验的演化过程，书中的插图和设计案例即是不同时期的教学过程的部分呈现，参与这些教学实验的同学名单如下：

中国美术学院2020级研究生：
刘旭超、苏海璐、张涵、邵佩琳、竺央、褚杰

中国美术学院2017级城市设计系本科生：
戴源、陈沛泓、杨涵、王泽美、胡艺璇、陈思含、韩逸宁、赵安娜、蔡依婷、徐子莹、吴铠辰、熊洁、程梓瑜、刘怡雯、钱雨瑶、林琪翔、郭毅扬、纪欣园、何晔、孙馨蕾、梁沅沅、何子畅

中国美术学院2016级城市设计系本科生：
章雪滢、宋禹正、侯天艺、兰天、李红阳、张强、谢雨暄、蒋雨洋、谢韫灵、谢天怡、余奕影、何琛宇、王革、吴筱恬、汤佳、李千翔、范雨赞、彭若清

中国美术学院2012级城市设计系本科生：
俞嘉懿、王格格、赵娅宁、朱曦、陈砚风、郁修懿、黄栎元、张露、杨家豪、马彦京、陈奕琳、周鑫、李倩云、衣伟